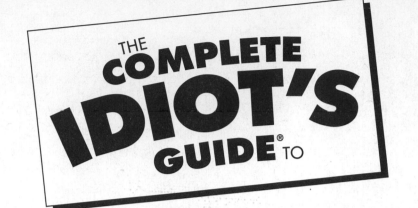

THE COMPLETE IDIOT'S GUIDE® TO

Understanding Intelligent Design

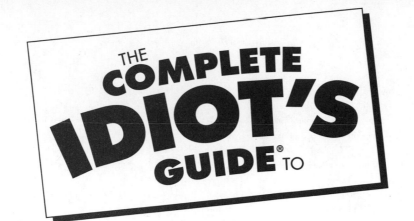

THE COMPLETE IDIOT'S GUIDE TO

Understanding Intelligent Design

by Christopher Carlisle, M.Div.,
with W. Thomas Smith, Jr.

ALPHA

A member of Penguin Group (USA) Inc.

For Ashley, Caddy, and Will, who give me purpose.

ALPHA BOOKS

Published by the Penguin Group

Penguin Group (USA) Inc., 375 Hudson Street, New York, New York 10014, U.S.A.

Penguin Group (Canada), 10 Alcorn Avenue, Toronto, Ontario, Canada M4V 3B2 (a division of Pearson Penguin Canada Inc.)

Penguin Books Ltd, 80 Strand, London WC2R 0RL, England

Penguin Ireland, 25 St Stephen's Green, Dublin 2, Ireland (a division of Penguin Books Ltd)

Penguin Group (Australia), 250 Camberwell Road, Camberwell, Victoria 3124, Australia (a division of Pearson Australia Group Pty Ltd)

Penguin Books India Pvt Ltd, 11 Community Centre, Panchsheel Park, New Delhi—110 017, India

Penguin Group (NZ), cnr Airborne and Rosedale Roads, Albany, Auckland 1310, New Zealand (a division of Pearson New Zealand Ltd)

Penguin Books (South Africa) (Pty) Ltd, 24 Sturdee Avenue, Rosebank, Johannesburg 2196, South Africa

Penguin Books Ltd, Registered Offices: 80 Strand, London WC2R 0RL, England

Copyright © 2006 by Christopher Carlisle

Publisher: *Marie Butler-Knight*
Editorial Director/Acquiring Editor: *Mike Sanders*
Managing Editor: *Billy Fields*
Development Editor: *Nancy D. Lewis*
Production Editor: *Megan Douglass*
Copy Editor: *Keith Cline*
Cover/Book Designers: *Kurt Owens/Trina Wurst*
Indexer: *Julie Bess*
Layout: *Eric S. Miller*
Proofreader: *Aaron Black*

Contents at a Glance

Contents

10 Mainstream Biology   105

Part 4: Intelligent Design Applications   117

11 Intelligent Design Physics   119

Introduction

Intelligent Design—or the idea that a cosmic "intelligent" entity designed the universe—is an idea that has existed for centuries: much of the evidence for that designer being found in the complexity of living organisms.

For most of recorded history, much less controversy surrounded the idea. Science and religion were generally deemed compatible fields of knowledge. Design was presumed to be the nature of creation, and for many, God created the design.

Classical science accepted this presumption of design for thousands of years—at least until a man named Darwin proposed his radical idea about the origins of life. Charles Darwin's ideas essentially withdrew the previously accepted "designer" from the biological world which he now described by the theory of evolution. Science would continue to move forward, but without the role of a transcendent designer who justified the seemingly miraculous structures evident in the universe.

Oddly, however, as science advanced, new discoveries revealed complexities so seemingly designed that philosophers and scientists began to reconsider Darwin's view. In the 1980s, Intelligent Design arose to meet the scientific challenge that random mutation and natural selection could explain our organic world. It is this movement that asserts the presence of design remains in the physical order, and some within it are even moved to contend that creation shows there is a designer.

Thus the current debate, which we look at in detail.

Most opponents of Intelligent Design—primarily those within the mainstream scientific community firmly grounded in Darwinian theory—argue that modern Intelligent Design is nothing more than masked creationism: the biblical account of creation, which scientists believe has been thoroughly debunked.

Most proponents argue that Intelligent Design has nothing to do with "creationism"; and it has everything to do with ancient approaches to science in the modern day. Rather than new technological research diminishing the role of design, they believe that it more dramatically reveals a designer in the universe. We present the evidence on both sides of the debate.

You decide.

What You'll Learn In This Book

This book is divided into seven parts, each of which is designed (no pun intended) to help you understand the root causes, the history, the basic premises, the arguments, and the current scientific and nonscientific conversations surrounding the debate over Intelligent Design. We all have an idea of what we think ID is, but few of us have an accurate understanding of ID's particulars.

Part 1, "Hey, What's the Big ID?" explores the fundamentals of Intelligent Design, including the current controversy and debate between science and "ID." You discover answers to basic questions such as: What is Intelligent Design? How do we define science? And how does Intelligent Design differ from creationism?

Part 2, "The History of ID," details just that: the history of Intelligent Design over the past 2,500 years, and up through the current debate between the proponents of ID and the mainstream scientific community.

Part 3, "'Traditional Science' Applications," offers a nuts-and-bolts review of classical science's "big three"—physics, chemistry, and biology—and their impact on the current debate over Intelligent Design.

Part 4, "Intelligent Design Applications," explains how Intelligent Design relates to science's "big three" and why ID proponents argue there must be a "designer" or, for most, a God behind—or beyond—the physical world.

Part 5, "Darwin's Meaning of Life," tells the story of Charles Darwin, the nineteenth-century British naturalist who not only gave us his theory of evolution by way of natural selection and random mutations, but also changed the existing scientific paradigm, making science less willing to accept the concept of modern Intelligent Design.

Part 6, "'Intelligent' Meaning of Life," looks at public perceptions of Intelligent Design and what ID proponents argue may well be a more adequate explanation of life's origins than that which has been adopted by present-day evolutionists.

Part 7, "Is a Resolution Possible?" asks—and attempts to answer—the difficult questions regarding the future of Intelligent Design, and why scientists, theologians, and ID proponents must come to some level of mutual respect, understanding, and common ground.

At the end of the book, you'll find helpful appendixes.

Extras

As you move from chapter to chapter, you'll notice little tidbits of fascinating information that will not only help you better understand Intelligent Design, but will shed light on the more difficult concepts and lend perspective to the ID debate.

Evolutionary Revelations

These boxes expound on historical snippets and interesting comments from experts that few people are aware of beyond the scientific, religious, and Intelligent Design communities.

Fact or Faith

Here you find explanations as to whether a given particular of ID is based on hard science, a theory, or a long-standing faith-based belief.

Blinded By Science

These boxes shed light on what readers might consider to be either controversial or difficult-to-understand science as it relates to evolution, creationism, or ID.

def•i•ni•tion

These are simply definitions, explanations, and elaborations.

Acknowledgments

As with any significant project, the writing of this book was not accomplished alone, and I have many to thank for many reasons. From the beginning of the marathon, they are Harris Pastides, for having first recommended me to do it, along with Patricia, for their cherished, faithful friendship; my talented co-author, W. Thomas Smith Jr., who put up with my literary eccentricities as a partner in crime; acquisitions editor Paul Dinas, whose wisdom insightfully guided the initial construction of the book; Steve Goodwin, whose gift as a teacher was richly utilized in developing the scientific ideas; and Dick Teresi, whose friendship, wit, and ability to write is an inspiration.

For keeping the home fires burning while I was sequestered away at the computer, I have Lonce and Julia Sandy-Bailey, and Merle Ryan to thank. Thanks also go to Dave Thom for his supportive advice and resource suggestions, and Craig Nicolson for his capacity for mundane examples. And without the support of Bishop Scruton, Bill Coyne, and Bruce Rockwell, the book could not have been written.

For the bowls full of conversation and red wine, I thank Don Fisher, whom I will miss as my next-door neighbor. I will always be indebted to the "Boys' Club": Mich Zeman, and his uncanny ability to make his presence known all the way from New York; Sid Poritz, whose deep friendship is an abiding presence in my life; and Lynn Klock, whose wife, Laura, deserves a medal for putting up with us both.

Last, and certainly not least, is my family. Thanks to Dad and his endless stream of articles from the *Times;* to Mom, for her endless stream of maternal support; to Claire, for being my forever mother-in-law; and to Cyn, Mark and Sue, Liz, and Tim and Karen—with the 10 terrific kids they have produced—for being the family they are.

Finally, my deepest love and respect go to Ashley and Andy; Caddy, Roberto, and Ethan; and Will, who are as much evidence of divine design as I will ever need.

Trademarks

All terms mentioned in this book that are known to be or are suspected of being trademarks or service marks have been appropriately capitalized. Alpha Books and Penguin Group (USA) Inc. cannot attest to the accuracy of this information. Use of a term in this book should not be regarded as affecting the validity of any trademark or service mark.

Part 1

Hey, What's the Big ID?

The current competition between classical science and Intelligent Design may well be one of the most important debates in our culture's history. It is important to our grasp of the nature of science, religion, and knowledge overall.

So to engage in this debate with some semblance of understanding, we must first answer the three big questions: What is Intelligent Design? What is science? And how does Intelligent Design differ from biblical creationism?

Chapter 1

What Is Intelligent Design?

In This Chapter

- ◆ The debate over Intelligent Design
- ◆ Origins of the Intelligent Design idea
- ◆ "Masked creationism" or scientific inquiry?

Intelligent Design—or the idea that an intelligent agent (with an intellect greater than anything we as humans can fathom) designed the universe we live in—may well be one of the most polarizing debates of the twenty-first century. And for good reason because the two camps within the debate are separated; not only by what may or may not be "pure science" as we understand it, but by the question of whether or not there is a God. And if so, did God create the universe—including us?

The champions or proponents of Intelligent Design (which we often refer to in this book as ID) argue that the basic structure of our universe (more specifically the building blocks of the earth's myriad life forms) are far too detailed, beautiful, mathematically formulated, and deeply complex to have begun by a chance meeting of two or more elements. ID poses further questions: Where did the two or more elements come from? Did they just happen? Or was there conscious design behind them? ID proponents

Evolutionary Revelations

A November 2004 CBS News poll revealed that more than half of Americans surveyed believed that God created human beings in their present form.

argue that a designer—whether it was God, aliens traveling from another universe, or someone or something from some other dimension—had to have been involved. But if our design was the work of ultra-intelligent space aliens, who or what designed the aliens themselves?

The idea that there exists an Intelligent Designer is viewed by opponents of ID as nothing more than "masked creationism"—the idea that God created the universe. It is simply not provable science, they say. Moreover, they argue, the quest to prove the viability of an Intelligent Designer is fraught with contradictory problems beyond the fact that we cannot physically see the designer, or "God."

"For the advocates of intelligent design, the loveliness of nature is a second-class road to truth," writes author-journalist Garret Keizer in a March 1, 2006 article for the *Houston Chronicle*. "It is 'merely' aesthetic. In that regard, one notices that there is no campaign afoot to teach 'divine inspiration' as the basis for the sacred works of Fra Angelico and Bach. 'That's next,' you say, and maybe it is next. The point here is that it wasn't first, and it wasn't first for a very good reason."

Certainly the debate is hot. It is not going away. And if we are to understand it, we must examine its historic roots, current arguments, and evolving theories.

Defining Intelligent Design

Let's face it, these days it is almost impossible to leaf through a newspaper or a magazine without seeing some reference to the debate over Intelligent Design.

Fact or Faith

It doesn't really matter what position you take. This is a hot topic. It is in the newspapers every day. It is talked about in colleges and universities across the country.

—The Rev. Paul Veit as quoted by Hilary Snow in "Ascent of man, or God's plan?" *The State Port Pilot*, February 22, 2006

In most cases, we read or hear about it discussed in the context of Darwinian evolution, perhaps because the tenets of evolution fly directly in the face of Intelligent Design. Because of its importance in the debate, we devote a portion of the book to the understanding of evolution. But as we shall see, Intelligent Design has a long and fascinating history that applies not only to evolution and evolutionary biology, but to all of the other major scientific disciplines as well.

So how do we define it? The Seattle, Washington-based Discovery Institute (a think tank for Intelligent

Design comprised of some of the leading voices in the debate) defines modern Intelligent Design as a theory that some things are "best explained by an intelligent cause," though not necessarily God.

Fact or Faith

The scientific theory of Intelligent Design holds that certain features of the universe and of living things are best explained by an intelligent cause, not an undirected process such as natural selection … Intelligent Design theory does not claim that science can determine the identity of the intelligent cause. Nor does it claim that the intelligent cause must be a "divine being" or a "higher power" or an "all-powerful force." All it proposes is that science can identify whether certain features of the natural world are the products of intelligence.

—Discovery Institute

Something "Caused" Us

It's easy to assume that Intelligent Design is a philosophical position, or more popularly, a religious position that strives to scientifically prove God. And insofar as biological Intelligent Design is concerned with the origins of life, it is destined to have religious implications. But in its purest form, Intelligent Design contends that the random processes described by evolutionary theory are inadequate to explain the complex biological structures found in organic life, such as the human eye or the blood clotting system, which will be looked at in Chapter 13. Therefore ID looks to "intelligent causes" to understand these complex structures, developing means of empirically detecting these causes over and against undirected natural causes.

William Dembski, a mathematician and Discovery Institute fellow, naturally perceives Intelligent Design through the lens of his own discipline. Dembski sees Intelligent Design as a theory of information that, when it is successfully carried out, can reliably indicate "intelligent causation." (We get more into such "causation" a bit later.)

"Intelligent design is therefore not the study of intelligent causes per se but of informational pathways induced by intelligent causes," he says. Like most proponents of Intelligent Design, Dembski is quick to point out that a specific creator or agent can never be identified by such a process. Calling it "theologically minimalist," he says, "[Intelligent Design] detects intelligence without speculating about the nature of the intelligence."

Dembski is not saying that God designed or created the universe, but that something did.

Looking Backward

Proponents of Intelligent Design see ID as a thoroughly scientific theory.

Unlike philosophical arguments for design, ID begins with the empirical evidence of the natural world. This may be radically different from what many of us imagine ID to be. Rather than beginning with a personal conviction about the existence of God and then looking for natural features that substantiate—or support—this belief that God exists, Intelligent Design goes first to the material features that it deems worthy of exploration. It investigates these material features with scientific tools to determine the presence of design.

When "intelligent design" is indicated to ID proponents, the scientific question becomes: how was it produced? Dembski calls this process *reverse engineering*.

def•i•ni•tion

Reverse engineering is a relatively simple concept in which—instead of working from the source to the object to prove a theory—a design-theorist attempts to work back from a given designed-object to speculate about the various ways that it might have been made.

The Violin Model

Dembski likens the process to exploring how a Stradivarius violin was created, invoking a sense of "mystery" that isn't found in mainstream science. Instead of concluding right off the bat that Italian violin maker Antonio Stradivari was the principle designer behind the Stradivarius violin, ID looks first at the violin itself and works backward from the instrument to discover the man who made it.

One might logically ask: if an ID scientist concludes that an object was "designed," what is there left to do? ID proponents argue that after Intelligent Design is determined, the party has just begun, and scientists can bring investigative tools to bear in achieving greater knowledge. The thinking goes that because scientists themselves are creatures of Intelligent Design, "Intelligent Design is one intelligence determining what another intelligence has done," says Dembski.

Always an I-Design Idea

Arguments for design in nature began with the classical Greeks and today provide some of the underpinnings of modern Intelligent Design. But the actual term *Intelligent Design* emerged in the nineteenth century when it was used in an address several years after Darwin's *The Origin of Species* was published.

Speaking to the British Association for the Advancement of Science, botanist George James Allman said, "No physical hypothesis founded on any indisputable fact has yet explained the origin of the primordial protoplasm, and above all, of its marvelous properties, which render evolution possible—in heredity and in adaptivity, for these properties are the cause and not the effect of evolution. For the cause of this cause we have sought in vain among the physical forces which surround us, until we are at last compelled to rest upon an independent volition, a far-seeing intelligent design."

God Temporarily Removed?

Darwinian evolution gets most of the credit for taking God out of science. If so removing God is a crime to many of us in the twentieth-first century, it was a felony to most in nineteenth-century Victorian religious culture. Add to the mix the especially religious character of culture in America, and at the very least Darwin ensured heated debate across the sea.

Although there were frequent nineteenth-century criticisms of evolution from the church pulpit, there were few such scientific counterarguments until the 1920s. It was then that the conflict ceased to be a simple religious criticism by the church and became a debate between (mostly) Christians who weighed in with the authority of science. Most modern scientists would judge these arguments as at best "junk" or "pseudo-science," because in retrospect they seem like a thinly veiled disguise of religion, indeed, Christian faith.

"In the Beginning ..."

As we shall see, creationism does in fact figure into the Intelligent Design debate. And in following chapters we look at creationism over and against Intelligent Design. But for now, "creationism" will be treated as ID's primitive beginnings that led us to the current conflict.

In its original form, creationism was based upon theology alone, contending the origins of the universe happened according to the biblical accounts.

Fact or Faith

In the beginning God created the heavens and the earth. The earth was without form and void, and darkness was upon the face of the deep; and the Spirit of God was moving over the face of the waters.

—Genesis 1:1–2 (Revised Standard Version)

The Hebrew scriptures, or "the Old Testament" (to Christians), contain the creation story. The Bible opens with God making the universe out of nothing.

From there, the Bible describes the way the universe as we know it was created: the heavens, the earth, vegetation, stars, animals, and human beings all come into being to receive God's blessing as part of "his" dominion.

As one reads the account, the particulars of how the universe was made become secondary to the overwhelming point: it was God who made it.

In Genesis 2:1–3, the story continues, "Thus the heavens and the earth were finished, and all the host of them. And on the seventh day God finished his work, which He had done, and he rested on the seventh day from all His work, which He had done. So God blessed the seventh day and hallowed it, because on it God rested from all his work which He had done in creation."

In all likelihood, a modern editor would mark up this account as redundant. But these several repetitions underscore the message that God's power was established in creation and that creation itself stands as a living manifestation of God.

Evolutionary Revelations

One of the problems with basing one's understanding of creation on "the biblical account" is that there are two accounts: the first account appears in the first chapter of Genesis, wherein humans are created five days after the earth was made. The second account appears in the second chapter of Genesis, wherein humans are created directly out of the earth.

For some Christians, the biblical creation story is at best a "metaphor"—a myth that isn't rooted in fact—but for creationists it stands as a foundational assumption of their very faith.

Blinded By Science

The account of the great flood during the time of Noah, which figures heavily in the creationist movement, is considered dubious by many contemporary geologists. That said, it is interesting to note that many societies and cultures—disconnected from one another for centuries—worldwide, hold to a tradition that a great flood of biblical proportions once covered the entire Earth. Additionally, some scientists believe there is historical evidence of such significant flooding in the Mediterranean world.

A "Young Earth Creationist"

In the tradition of accepting the creation story in Genesis as fact rather than fiction, George McCready Price got the ball rolling in the early 1900s toward a revival of the biblical account as an empirical truth.

As a "young-Earth creationist," Price rejected the mounting scientific evidence in the field of geology that the world was a whole lot older then the traditional biblical account. According to the Genesis creation story, the world was created in six days; by the biblical math, creation must have occurred between six and ten thousand years ago.

Evolutionary Revelations

Native Canadian George McCready Price's best known book was *The New Geology*, a college textbook published in 1923, which made a number of argumentative cases against many of the key points in Charles Darwin's theory of evolution. Price's book and other works began to achieve notoriety during the famous Scopes trial, two years later, when William Jennings Bryan drew on the research of Price to support his own arguments.

Price also was a Seventh Day Adventist, a member of a small religious community led by "prophetess" Ellen G. White. In one of her visions, Ellen White claimed to have witnessed the creation, which occurred in a calendar week.

It appears Price wanted it both ways: he wanted "proof" of a young Earth, but a young Earth that was finally unprovable by science. Like creationists since, he asserted that the truth of creation lay beyond science.

Price would write, "… Creation is inscrutable; we can never hope to know just how it was accomplished; we cannot expect to know the process or the details, for we have nothing with which to measure it. The one essential thing in the doctrine of Creation is that the origin of our world and of the things upon it came about at some period in the past by a direct and unusual manifestation of divine power …."

It is important to note that twentieth-century creationism was not a monolithic movement. A vast array of opinions about how creation happened made for anything but a unified view; in fact, the movement's many competing statements of belief describe a conflict-ridden history. The one thing that can be said, for the wide diversity of Genesis interpretations, is that to creationists, the Bible was the ultimate authority concerning how the world was made.

Contemporary Creationism and ID

We will see that creationism has continued on into the twenty-first century. And we will see how creationism and Intelligent Design have been presumed to be one and the same. But like every history, the historical strands are rarely discrete or clear; they are an interwoven tapestry of partial perspectives held by finite human beings.

The term Intelligent Design reared its modern head in 1988. That year, Charles Thaxton, editor of the book, *Of Pandas and People: The Central Question of Biological Origins*, used it in a conference presentation, and then in his forthcoming book.

Phillip Johnson, considered a founding father of modern Intelligent Design, gave it profile in his book *Darwin on Trial*, which was published in 1991. As an alternative to the mainstream scientific assumption that God was not a part of science, *Intelligent Design* became a scientific term that incorporated religious understanding.

ID's Think Tank

It was during the 1990s that the Seattle-based Discovery Institute was founded. Considered Intelligent Design's principal think tank for ID public policy strategy, the Discovery Institute now works to influence legislation, education, and the perception of ID in the culture. Prominent advocates are counted among the "fellows" of the institute, and its Center for Science and Culture has successfully guided the growing movement in America.

Fact or Faith

Board members and fellows of Intelligent Design's Discovery Institute hail from a wide variety of religious and philosophical backgrounds: mainline Protestant, Roman Catholic, Eastern Orthodox, Jewish, and agnostic.

Unlike the strictly theological beginnings that undergirded creationism, many of the leading voices in the current ID movement are trained scientists: biochemists, biologists, physicists, and mathematicians all appeal to "scientific method" in their contention that science alone cannot explain the universe, or human life. To the extent these scientists hold religious convictions, they claim to exclude them from their science. The design they strive to see is not a product of their faith, but of their scientific inquiry.

A Hidden Religious Agenda?

Critics of Intelligent Design claim ID has hidden or "masked" religious motives. They contend the ID movement is an effort to push a religious agenda on the culture—particularly in colleges, in the public schools, and in the political arena.

In response, leading voices in the ID movement assert that scientific orthodoxy has demanded a naturalistic agenda that precludes the supernatural. They contend that mainstream science, as it is now defined, has an atheistic agenda, and all that they are asking is an equal hearing in the schools, the courts, and the culture.

Undoubtedly some of the confusion between creationism and Intelligent Design came by way of Phillip Johnson shortly after the movement got going. Johnson stated that the goal of Intelligent Design was to cast creationism as "scientific." But now the Discovery Institute makes it clear that in no way is Intelligent Design "creationism"—a substantive distinction.

A Growing Movement

Intelligent Design grows in strength and influence as it competes for cultural attention, and its critics are becoming as outspoken as proponents of Intelligent Design themselves. Among them is Barbara Forrest, considered an "expert" on the movement, who believes ID is guilty of deliberately masking what she thinks of as a religious agenda. Insisting that ID is biblically based, she fears an ulterior motive: that the movement is challenging the First Amendment—the separation of church and state. She has written that the ID movement's activities "betray an aggressive, systematic agenda for promoting not only Intelligent Design creationism, but the religious view that undergirds it."

Like every historical movement, Intelligent Design has changed and evolved over time. Whatever the motives—whether political strategy, or a transforming self-understanding—ID naturally appeals to those with religious sympathies. Yet as we shall see, ID is not content with biblical authority alone in its quest for understanding the relationship between God and the physical creation. ID attempts to use science as a way to reveal the presence of design which for some is simply seen as an "intelligent agent," and for others, as a sign of God.

The Least You Need to Know

- The debate over ID may well be one of the most polarizing issues of our time.

- Proponents of ID believe the universe is too complex not to have been designed.

- Opponents of ID believe Intelligent Design cannot be proven, and thus is not science.

- The concept of an "intelligent agent" designing the universe has been an idea for centuries.

- ID opponents argue that modern ID is "masked creationism."

- ID proponents come from many disciplines and several faiths.

What's the Scientific Problem?

In This Chapter

- The notion of science?
- What or who is God
- Nonoverlapping magisteria
- The conflict

When many people—who are coming into the debate cold—think of Intelligent Design, they wonder why proponents of ID believe ID to be science. Isn't ID, after all, a theological discipline? Yes and no: yes, in the sense that there is certainly plenty of room for God in ID. No, in the sense that ID does not necessitate that the Intelligent Designer be God, nor that God be the focus of the inquiry. For some ID proponents, the nature of the particular design itself is of greatest scientific interest.

But whether or not we see ID as science depends upon how we look at science, and what we consider science actually to be.

So we are about to take a look at the nature of both science and theology, and why—in the minds of many mainstream scientists—the idea of

Intelligent Design is believed to inappropriately integrate scientific inquiry with religious faith. We will also look at why so many scientists believe that science and faith should be kept separate. The fact is it wasn't always this way.

Let's consider the conflict.

What Is Science?

How do we answer the question, "what is science?" Many people may think the answer is obvious; after all—isn't science a fundamental staple of every high school education? And isn't science one of the great engines of American culture, giving us unprecedented technology and incredible insights into the nature of life? Isn't science, just, well, science?

It may not be as simple as that. Science is not always easy to define, and throughout history there have been a variety of interpretations as to just what science really is. Not always being able to define science puts us in good company.

Teresi's Lost Discoveries

In his book, *Lost Discoveries*, noted science writer Dick Teresi attempted to lead off with a good, solid definition of science. But what he realized is not only that "science" is difficult to define but there may never have been a watertight definition of science.

 Fact or Faith

There is no good definition of science. [The American Association for the Advancement of Science], for example, does not have one. After many trials, the American Physics Society [for physicists] finally decided upon a definition. The APS found that if the definition was too broad, pseudosciences like astrology could sneak in; too tight, and things such as string theory, evolutionary biology, and even astronomy could be excluded.

—Dick Teresi, *Lost Discoveries*, 2002

Nevertheless, Teresi distilled a general understanding of science. He says science is "a logical and systematic study of nature and the physical world." He goes on to remind us that science normally involves theorizing and experimentation.

The Orange Juice Test

So let's consider, experimentation: we might remember from high school that the "scientific method" generally consists of six elements: hypothesis, apparatus, method, observation, results, and conclusion.

Eyes starting to glaze over? Bear with us for a moment while we make it simple by looking at these six elements as we apply them to a glass of orange juice.

For instance, as scientists we might "hypothesize" that orange juice is acidic and that we want to scientifically prove this hypothesis. We might do this by bringing together three things—the glass, the orange juice, and some litmus paper (which turns color in acidic solutions). These three things together form an "apparatus" to test the hypothesis. Our "method" might be to dip the paper in the glass of juice for 30 seconds. Then we would remove the paper from the juice and "observe" whether it turns color. If, barring any unforeseen factors, the "result" is colored paper, we might logically "conclude" that orange juice is acidic.

Although this is a simple example of the scientific method, it is followed in most of science. From physics to chemistry to biology, and their various subdisciplines, scientists go about observing the world by theorizing and experimentation. And whatever prior beliefs scientists might bring to the lab, the proof is in the juice—so to speak— freeing objective truth from prejudice and allowing the natural world to speak.

Science as "Empirical"

So what is scientific knowledge? It is the knowledge that comes from what's known as "empirical" evidence. In other words, from the observable stuff of life, which must be tested to verify any hypothesis we might entertain about it.

Now this may cause you to say, "Well, of course. How else could we go about knowing that something is really true?" But we must understand that before the rise of the scientific method we have today, there were other modes of understanding that departed from the concrete experience that is now so prized by mainstream science.

The Greeks, for example, didn't like empirical investigation. They found experimenting messy and believed that it polluted the ultimate purity of knowledge. Only abstract "forms" of knowledge could embody truth, whereas the concrete world distorted it. Therefore Plato's abstract "idea" of a kitchen table was more "real" than any one rendered by muscle, screws, and wood.

Of course, this perspective seems foreign to us. After all, if the abstract is real, it is only real when it has implications for the material world (reminding us of the power that science has in the way we think and see the world). What truly matters to us in the twenty-first century is orange juice we can drink, rather than some abstract "form" of orange juice; knowledge that is real is knowledge that is physically useful to us.

Science as "Repeatable"

That scientific inquiry is repeatable seems equally self-evident. If we dip the litmus paper in the glass of juice, and it indicates acidity, will we do it again to see whether we achieve the same result? Probably, yes (assuming the conditions are identical). In fact, repeatability is generally held as a prerequisite to science.

By retesting the hypothesis, not only can we check whether the experiment was "done right," but we can begin to entertain the notion that there is a scientific truth that works beyond the initial experimental moment. If the juice is truly acidic, it will color the paper again, indicating a characteristic which we can presume will always show itself, thereby vindicating our hypothesis about it.

Science as "Falsifiable"

As previously mentioned, defining science has long been a difficult problem. The "principle of verifiability" was proposed in the early twentieth century as a feature of all legitimate scientific inquiry, stating that unless a theory is able to be verified, it can't be considered scientific. In other words, if I can't verify that my orange juice is acidic, it can't be deemed a scientific truth. In short, nothing can be deemed scientifically true unless it can be proven.

Or can it?

Austrian-British philosopher of science Karl Popper criticized this verifying principle, saying that nothing can ever be positively verified. Repeated experimental results, he said, increase the chance that a hypothesis is true, but do not confirm it absolutely. In fact, nothing can.

For instance, gravity is a theory that we accept as true because it has always worked for us, but we can't know with absolute certainty that one day an apple won't fall up when we release it.

So Popper argued, falsification (now widely known as falsifiability) is a preferable criterion for science. Considering our orange juice example, Popper would have probably said that because our orange juice test is free to show that orange juice is either acidic or nonacidic, it qualifies as scientific inquiry. If the test were not free to show that the OJ were nonacidic, it would have to be judged scientifically meaningless.

This brings us to a natural next question: what cannot be "falsified"?

Fact or Faith

One problem with "falsification" as a criterion for science is that not all credible scientific theories can be falsified. Superstring theory, for instance, a current "grand unified theory" of physics being worked on by hundreds of physicists around the world, would require a particle accelerator 10 light years in diameter to falsify it. And that's probably not going to happen.

Studying God

The subject of theology—or studying God—long precedes such formal science. Indeed, the modern-day academy—our colleges and universities—grew directly out of the church. In the Middle Ages, they were one and the same. Therefore, to study math was to "go to church."

It would be easy to assume that there is little rigor in this discipline of "studying God" that we call theology. On TV we rarely see thoughtful theologians developing systematic thought; it is the newsworthy, simplistic religious sound bytes that normally get the play. But many of the greatest minds in academic history were in fact theologians, and most of our celebrated universities began as religious institutions. Look at Harvard, Yale, and Princeton: they all prepared young men for the ministry before they branched out to support the diversity of studies they explore today.

Miracles and the Spirit World

As we shall see, the wall between the natural and the supernatural worlds was not as absolute as it has become in our modern scientific culture. Supernatural assumptions (spirits, souls, miracles, the afterlife, the abstract, the unexplained, etc.) about the world and the universe permeated western culture through the Middle Ages, critically impacting the arts, education, and the rhythm of everyday life. Yet in our world the current wall that exists between the natural and the supernatural has created a cultural point of view from which supernatural things often seem to be "unreal" additions to a more "real" natural world.

Even persons of some religious faith are inclined to separate the two: more often than not, people look to God for answers to spiritual questions and to science for answers about the world. Often people in the modern world are apt to simply say each has a different focus:—the heavens, as opposed to the earth. Yet this need not mean there must be conflict between the two: just a logical separation, and a clear understanding that the "supernatural," by its very definition, simply transcends or goes beyond the natural world.

Religion Not "Empirical," "Repeatable," "Falsifiable"

We already have said that science depends upon three important criteria: that a given body of knowledge be "empirical," "repeatable," and "falsifiable," to be qualified as scientific knowledge or scientific "fact."

Some might say that religion fails on all three counts, or that it isn't testable at all. Others would argue that it is testable, but by an entirely different set of standards.

Most modern scientists would heartily agree and agree on scientific grounds. If the stories of the Bible are literally true, then they are one-time historical events. If they are not literally true, then by their very nature they are not empirical.

In any event, how could we possibly "falsify" the existence of God? How could we test to see whether God does not exist in order to prove that God did exist? There is simply no glass of juice or litmus paper to vindicate one's faith in God.

Blinded By Science

One reason the term "falsifiability" may be hard to get one's head around is that it makes us think in terms of the negative, which—for most of us—is an unusual or awkward way of thinking about things. Why would we want to know what isn't true rather than what is true? We tend to think in the positive—what is, rather than what is not.

Miracles Defy Modern Science

So we come to "miracles." If there is any realm where the relationship between God and science is strained, it is in the Biblical stories of the miracles so venerated by the church. To many mainstream scientists, miracles seem scientifically absurd—an offense to the scientific method and the truths of the natural world. To the church, they are sacred realities whereby the supernatural world interacts with the natural world.

Yet the modern-day assumption that miracles are an affront to scientific inquiry is very different from that of the Judeo-Christian world 2,000 years ago. Whereas one might presently sense that science and religion are two competing kinds of faith, it would appear that back then miracles lived more peaceably with the natural world.

In our scientific age, miracles seem to stand as a problem between science and religion. Perceived by "nonbelievers" as a means to "prove God" over and against other faiths, secularism, and atheism, miracles seem to defy the very tenets of science, and in so doing, the very "nature" of the material world.

Not empirical, repeatable, or falsifiable, miracles have come to represent a primary reason why science and religion are seen as never to be reconciled.

Fact or Faith _____

According to the traditional view, a miracle is "a sensible fact" (opus sensibile) produced by the special intervention of God for a religious end, transcending the normal order of things usually termed the Law of Nature.

—F. L. Cross and E. A. Livingstone, *The Oxford Dictionary of the Christian Church,* 1978

Not Overlapping the Two

Nonoverlapping magisteria is a fancy term coined by evolutionary biologist Stephen Jay Gould. All it really means is that science and religion need to be kept separate. As two distinct realms of knowledge, they cannot be mixed without distorting the other; and the way they go about ascertaining truth requires two entirely different operations.

As a scientist, Gould is aiming at Intelligent Design. Specifically, he believes that Intelligent Design is in reality religious inquiry, and that the very nature of the scientific method is therefore being violated. Science requires that all philosophical prejudice (particularly religious belief) be kept out of the objective process of ascertaining scientific truth.

> **Evolutionary Revelations**
>
> Whereas today Christians tend to see miracles as public signs of the truth of God's existence, we are told in the gospels that when Jesus performed miracles it was his practice to admonish others to tell no one.

Science and Religion's Unnecessary Friction

Gould says religious inquiry should be equally protected from science and that this would allow the two to enjoy a peaceful coexistence. His point is a defensible one: there should be no friction between science and religion precisely because each has its own goal, and its own unique way of doing business. It was this distinction that led Gould to write his now oft-quoted article, "Nonoverlapping magisteria."

> ### Evolutionary Revelations
>
> The net of science covers the empirical universe: what is it made of (fact) and why does it work this way (theory). The net of religion extends over questions of moral meaning and value. These two magisteria do not overlap, nor do they encompass all inquiry (consider, for starters, the magisterium of art and the meaning of beauty). To cite the arch clichés, we get the age of rocks, and religion retains the rock of ages; we study how the heavens go, and they determine how to go to heaven.
> —Stephen Jay Gould, *Natural History*, March 1997

Gould points out that the Catholic Church agrees that religious inquiry and science may peacefully coexist. Specifically, Pope Pius XII stated in his 1950 encyclical "Humani Genesis" that Catholics were free to believe in evolution as long as they believed that creation was divine and that human beings possessed souls. Pope John Paul reasserted the position of nonoverlapping magisteria, stating, "In his encyclical Humani Genesis, my predecessor Pius XII has already stated that there was no opposition between evolution and the doctrine of faith about man and his vocation."

Gould calls for "mutual humility" to be exercised between the magisterium of science and the magisterium of religion. He argues that the question of a "human soul," for example, lies outside of the bounds of scientific inquiry and shouldn't be pursued by science. In the same way, Gould believes scientific conjecture lies beyond the domain of religion, and that such movements as Intelligent Design violate this separation.

Science Need Not Be Violated

The current polarization between science and religion over the issue of Intelligent Design can lead us to believe it is one or the other—either science or religious faith. But undoubtedly you know people who believe in God and who also believe in science. Indeed, some of these people are scientists themselves, who engage both every day. So one might ask, how do they understand this seemingly conflict-ridden relationship?

For many scientists, the relationship should be one of absolute separation. Later we examine the role of "theistic evolutionists" (evolutionary biologists for whom God and evolution are quite compatible). In any case, most mainstream theistic scientists reject Intelligent Design's assumption that metaphysical conjecture can be part of science.

"Nonintelligent" Design

Let's state the obvious here: mainstream science and Intelligent Design are at odds with one another. In fact, one might argue, the debate between the two is as contentious as any in our current culture. And why not? Politically there's a lot at stake for both sides—or so they believe.

Although the debate is most intense around evolution, the question of Intelligent Design bears on every scientific discipline, with far-reaching implications, including the following:

- Where do we come from?

- Where are we going?

- What is the meaning of this life?

- Is there life beyond this world?

- What is the nature of God?

- Can God's existence be proven?

All are personally important questions; therefore it's no wonder that many Americans are so deeply invested in the issue. As we'll see, there's a marked disparity between scientists' common atheism and the beliefs of the typical person on the street who is asking significantly more religious questions.

"Intelligent" Mechanisms Explained Within Nature

For many scientists, Intelligent Design is an illusion. In their opinion, material creation does not require an outside "agent" or "God."

From the beginnings of the universe, to the origins of biological life, to the constancy of the earth's orbit around the sun, science supplies answers that successfully explain the function of the physical world.

Using physics, chemistry, and biology—as well as their burgeoning subdisciplines—life can be understood by natural laws which seem embedded in the universe. In our scientific age we can be free of "superstitious" explanations that may impede what we can actually know.

No Such Thing as Ghosts?

It is important to remember that most mainstream scientists—at the very least—question the existence of God. In fact, science questions everything. This may help to explain the common suspicions within the scientific community of anything that hints of the transcendent.

For many neo-Darwinists—modern-day followers of Darwinian evolution (we focus on Darwin himself and Darwinian evolution later)—there is no such thing as "spirit," or anything else that transcends the physical world.

It is such metaphysical claims that bothered Stephen Jay Gould. Here's why: if scientists have no right to claim that God can be proven in the natural world, neither do they have the right to proclaim that God does not exist. But as we'll see, they do: for many neo-Darwinists, there's no such thing as "spirit"—defying the very condemnation they make of Intelligent Design.

On the opposite end of the spectrum stand proponents of Intelligent Design. Not only do they personally believe in an "intelligent agent" or in "God" they believe it is scientifically indicated in the structures of the physical world. Thus, the conflict isn't simply about the content of people's beliefs—it is a conflict about the nature of science itself and the very way that human beings can know God.

The Least You Need to Know

- Most mainstream scientists believe science and faith should be kept separate.

- Though there is no pure definition of what science actually is, the consensus among scientists is that it must meet certain criteria: that such inquiry be "empirical," "repeatable," and "falsifiable."

- According to most relevant definitions, religious inquiry is not "empirical," "repeatable," or "falsifiable."

- The Roman Catholic Church has asserted that followers of the Roman Catholic faith are free to "believe in evolution as long as they believe that creation was divine and that human beings possess souls."

- Miracles are scientifically unexplainable events wherein the supernatural world interacts with the natural world.

- The question of Intelligent Design bears on every scientific discipline, with far-reaching implications.

Intelligent Design vs. Creationism

In This Chapter

- ◆ Creationism in the modern world
- ◆ Floodgates open
- ◆ Creationists deception or honest ID?
- ◆ The politics of it all

We've heard the argument from opponents of Intelligent Design that ID is actually "creationism" in disguise. In fact, much of the recent debate has centered on whether ID is "masked creationism." Though it is a controversial debate that will continue, it is important to understand the particulars of the disagreement between mainstream scientists and proponents of Intelligent Design. Wherever one ends up in the debate, it is clear that there are members of both camps who are thoughtful people who have distinguished themselves in their various fields of endeavor, and yet who substantively disagree with one another.

However, at this time in history, it may be that the greatest disagreement surrounding Intelligent Design has to do with whether ID and creationism are one and the same. ID supporters contend that both the philosophical assumptions and the way ID goes about its inquiry are fundamentally different from the assumptions and methodology of "creation scientists." We shall see that mainstream scientists have traditionally dismissed such a distinction as a deliberate tactic to appeal to those who are more scientifically inclined, while—at the same time—holding to the tenets of creationism.

The outcome of this debate has significant political implications. The conflict rages in American schools and in the courts—as America decides whether ID can be included as a part of public education. For if ID is deemed a form of religious expression, then clearly it may not be taught. However, if it is considered science, then it may be taught in the classroom as legitimate subject matter.

The Contemporary Take on Creationism

Creationism has never been a monolithic movement. Throughout its twentieth-century history, creationist organizations—each with its own unique belief system—have done theological battle with one another, and to this day the movement continues as a complex tapestry of religious beliefs. Indeed, in 1984 it was reported that there were at least 22 national creationist organizations, as well as 54 state and local groups. And they are not all working together.

Modern creationism really got going in the early 1960s. With the publication of their book *The Genesis Flood*, Henry Morris and John Whitcomb made a cultural splash. If the creationist movement had been in the background throughout the twentieth century, it came to the forefront with its bold assertion that the earth was made just as the Bible described.

The Bible, Literally

So what, you may ask, is the big difference between ID (wherein there is an Intelligent Designer) and creationism (wherein there is an Intelligent Creator—or God)?

Creationism begins with the Bible. For creationists, the first move is to understand how creation and material life fit into the Bible, rather than the other way around. It is therefore no coincidence that Biblical creationists are uniformly religious people.

Dr. Henry M. Morris, a hydraulic engineer and professor of hydrogeology, began to contemplate how the great flood described in Genesis could be scientifically understood. It was this question that spawned his writing of *The Genesis Flood*, called by

Harvard evolutionary biologist Stephen Jay Gould, "the founding document of the creationist movement."

Evolutionary Revelations

Dr. Henry Morris was one of the leading voices in the modern creationism movement until his death in early 2006. After growing up "religiously indifferent," Morris ultimately accepted the Bible "from Genesis to Revelation, as the infallible and literal word of God."

Morris founded the Creation Research Society, and he is considered in many circles to be the "father of modern creation science."

Now in its forty-fourth printing, *The Genesis Flood* articulates the creationist position that the Bible bears literal truth regardless of what science might say. Morris criticized the view that the Bible could be seen as a mythological text, calling such an assumption the beginning of an end whereby "every man becomes his own God."

Creationism, Young and Old

"Day-age" creationists are those who argue that each of the six days recounted in Genesis is actually an enormous, generally unspecified, stretch of time that far exceeds the 24 hours we normally think of as "a day." By lengthening the duration of "a day," day-age creationists are thereby able to explain how the Bible could attest to 6 days of creation in the face of the scientific evidence to the contrary. Jehovah's Witnesses stand as one example of such "old-Earth" creationism.

Gap creationists handle the science/faith problem somewhat differently: there is a gap, an undetermined lapse of time, between the first and second verses of Genesis. Following, "In the beginning God created the heavens and the earth," there is a temporal break before God moves "over the face of the waters," and finally declares, "Let there be light." This way, the 6-day creation can have occurred long after the origins of the universe, allowing for an "old" rather than a "young" Earth, and so similarly avoiding scientific refutation.

def•i•ni•tion

Gap creationism (also referred to as "restitution creationism" or "ruin-reconstruction") is a term that describes a fundamentalist Christian-based belief system in which the creation of the earth is believed to have taken far longer than 6 days. In gap theory, there is a "gap" in the 6 days that may have lasted thousands of years.

Then there are the "strict creationists." These are known as the "young-Earth" creationists who believe that creation happened precisely as described in Genesis. Thus the world was made in 6 days, and by the biblical generational accounting, between 6 and 10 thousand years ago. Unlike the "day-age" and "gap" creationists, who accommodate an old-Earth creation more consistent with geological research, the young-Earth creationists recognize little authority or influence beyond that of the Bible.

The Great and Terrible Flood

George McCready Price, an early founder of the American creationist movement, was a young-Earth creationist. In his 1923 book *The New Geology*, McCready explained that the world's current geological features are a product of the Biblical flood, which submerged all of life, save Noah, his family, and the animals on the ark. The presently visible sedimentary layers of the earth, he contended, are evidence of the flood; and its fossil deposits reveal the organic life which existed at the time of the deluge.

Using his hydraulic engineering training, Morris attacked contemporary geological research which argued for the existence of an old Earth. He rejected the "uniformitarian" geological position that the layers of sedimentary rock implied a multibillion years old planet, invoking "Biblical catastrophism" to explain how this enormous span of time could be collapsed to approximately 300 days. In so doing, the biblical account to which Morris was wedded could remain intact.

Fact or Faith _____

In the Biblical account of the great flood, God directs Noah to build an ark into which Noah, his family, and one male and female animal of every kind are to move before the entire Earth is flooded by God (in an act of divine retribution). It is interesting to note that "great flood" stories have been told for centuries, throughout the world, in societies that had no contact with one another.

Needless to say, there are scientific problems with such an explanation, and it provoked a dramatic response. In fact, to this day, Morris's bold contention still draws scientific response. In particular, evolutionary biologists are inspired to reply—perhaps more so than geologists.

Adapting, Not Evolving

The reason is clear: Morris's theory, however controversial, has evolutionary implications. For he argues that the evidence of increasingly complex mammals toward the uppermost sedimentary levels of the earth indicates not evolutionary development, but that these more advanced and mobile creatures were better able to struggle to the surface before they finally died from the flood. Rather than indicating evolutionary progress, Morris's theory allows for one distinct creation.

Even Stephen Jay Gould got into the act, poking fun at Morris's rejection of evolutionary logic in favor of able aquatics. "Surely, somewhere, at least one courageous trilobite would have paddled on valiantly (as its colleagues succumbed) and won a place in the upper strata," wrote Gould. "Surely, on some primordial beach, a man would have suffered a heart attack and been washed into the lower strata before intelligence had a chance to plot a temporary escape … No trilobite lies in the upper strata because they all perished 225 million years ago. No man kept lithified company with a dinosaur, because we were still 60 million years in the future when the last dinosaur perished."

That's a Lot of Water

With the goal of keeping the biblical account in tact, "Flood geologists" maintain that the whole Earth, including Mount Everest, was under water in the time of Noah. However, 1.1 billion cubic miles of water would have been required to fulfill the biblical description, whereas the earth's atmosphere has the capacity to hold only a fraction of the amount recounted by Genesis: "And the waters prevailed so mightily upon the earth that all the high mountains under the whole heaven were covered; the waters prevailed above the mountains, covering them fifteen cubits deep."

At the present time, the implications of flood geology for the age and origin of our physical world may be less important than its implications for evolution. For if the fact of such a flood allows for the belief in an Earth that is 10,000 years old, it also requires the sudden appearance of all life forms as described in Genesis. Thus the gradual evolutionary process by which mainstream scientists believe life came into being is brought into a dramatic question that defies the evidence.

Dueling Creationists

One might easily perceive creationism as a singular attack on mainstream science. However, the reality is that the movement bears a history marked by internal conflict and competing perspectives. It might more accurately be depicted as a complex, if not "scientific," struggle for human understanding. In fact some of the fiercest creationism conflicts have not been between creationists and mainstream scientists, but between creationists themselves.

In any event, creationism's conflict with mainstream science has been, and continues to be, real. What may be less obvious is creationism's conflict with Intelligent Design. As a forerunner of the ID movement, one might think that creationism would be proud of its prospering offspring. As we shall see however, a widening gap between their respective beliefs, coupled with the growing power of creationism's child, have served to underscore Intelligent Design's divergence from "the parent."

Neo-Creationism or Pure ID

The confusion between Intelligent Design and creationism may in part be generated by the insistence of mainstream scientists to blend the two. For some, they are identical enterprises; only the names have been changed (to jeopardize the innocent). For others, Intelligent Design is most appropriately named "neo-creationism"— essentially creationism that has been elaborated in modern terms to utilize the resources of modern science in making the creationist case. And there are those who recognize Intelligent Design as a movement unto itself, however they might disagree with its philosophical tenets or methodologies.

It's Not About Biblical Accounts

According to the proponents of Intelligent Design, ID does not presume the Bible. Unlike creationism, which looks to these scriptures for its ultimate authority, Intelligent Design purists claim empirical research as their singular source of knowledge.

Evolutionary Revelations
With its focus on empirical research, the Intelligent Design community rarely mentions or cites the Bible; and rarely is the name of "God" invoked as the conclusion of the presence of Intelligent Design.

ID looks for evidence of "intelligence": signs of "intentional" design. It is natural that mainstream science would suspect the "G word" is really what's at stake: after all, what else could they possibly mean by the presence of "intelligence"? Yet whether or not an ID scientist happens to believe in God, it is the contention of Intelligent Design proponents that this isn't part of their scientific program. In this regard it agrees with mainstream scientists that God cannot be known by science; however, this doesn't mean that the presence of an outside agent cannot be inferred.

Package Deal vs. Partial Truth

The methodological discrepancy between ID and creationism may best be illuminated by a recent exchange between two of their representatives. The now-deceased creationist Henry Morris and the leading ID theorist William Dembski engaged in a public discussion in which Dembski articulated what he felt ID was and was not able to assert. Responding to Morris's criticism that ID is "soft" creationism, Dembski conveyed what he believed to be the difference between the two.

"In inferring design from aspects of the world, we are always looking at finite arrangements of material objects and events involving them," said Dembski. "There is no way, logically speaking, to infer from such objects to an infinite, personal creator God … That's why intelligent design is not a biblical or religious doctrine. Morris is right that anyone except pure materialists can take refuge with intelligent design. This, however, should not be regarded as a bad thing. Creationism is a package deal, with a particular interpretation of the Bible being a part of the total package. Intelligent design, by contrast, is a partial truth, not the whole truth."

Although Morris was, and Dembski is, an avowed Christian, with regard to their modes of inquiry, they do not come out in the same place. In spite of common doctrinal beliefs, Morris had, and Dembski has, very different perceptions of how one can go about apprehending the empirical world. It is important to note that this not only affects their respective modes of inquiry; it also affects where they come out—their empirical conclusions. So Dembski would conclude, "… young-Earth creationism is … off by a few orders of magnitude in misestimating the age of the earth."

ID as Science

Asserting that Intelligent Design does not propagate religious belief, ID's proponents advocate its inclusion in public school curricula.

Evolutionary Revelations
Intelligent Design proponents contend that ID, by its refusal to name or infer the existence of "God," does not violate the First Amendment requiring the separation of church and state. They seek to ask questions they believe have gone unanswered by mainstream scientists and are asking for a fair hearing to the betterment of public education.

Much of the controversy about whether Intelligent Design should be taught in the public schools hinges on the issue of evolution as a competing philosophical perspective. Citing "holes" in the theory that are scientifically revealed by the theory of Intelligent Design, the Discovery Institute advocates ID's introduction as a way to better public education. Rather than calling for Intelligent Design to replace evolutionary theory, ID's think tank advocates the teaching of them both to the benefit of students' larger understanding.

Fact or Faith _____

Instead of mandating Intelligent Design, Discovery Institute recommends that states and school districts focus on teaching students more about evolutionary theory, including telling them about some of the theory's problems that have been discussed in peer-reviewed science journals. In other words, evolution should be taught as a scientific theory that is open to critical scrutiny, not as a sacred dogma that can't be questioned. We believe this is a common sense approach that will benefit students, teachers, and parents.

—Discovery Institute—Center for Science and Culture

The Politics of ID

It is interesting to note that the American Civil Liberties Union (ACLU) is once again in the midst of the creationist debate.

In 1925, the ACLU defended the right of a high school teacher to teach evolution in the classroom. Though it remains on the same side of the controversy—that is, against creationism—rather than working to include other perspectives, it is now working to exclude them.

Yet the commonality between these two ACLU positions is the insistence on separation of church and state. But for advocates of Intelligent Design, this smacks of political favoritism. It is ID's contention that scientific materialism implies an atheistic belief that systematically excludes other (religious) points of view that deserve a fair hearing.

Trouble in Ohio

In 2002, after months of debate, the Ohio State Board of Education unanimously adopted standards for the teaching of science which require that students understand the reasons "scientists continue to investigate and critically analyze aspects of evolutionary theory."

Its standard regarding evolution doesn't require the teaching of Intelligent Design, but only that the evidence for and against Darwin's theoretical framework be presented.

Dr. Jonathan Wells, a Discovery Institute fellow and author of *Icons of Evolution*, which analyzes "errors" in high school and college textbook presentations of Darwinian evolution, says biology curricula must be revised.

"Most biology textbooks continue to use outdated and discredited evidences like peppered moths and Haekel's embryos when it comes to their treatment of Darwinian theory." Wells said in a statement issued by the Discovery Institute on December 10, 2002. "It is critically important for school districts in Ohio to revise their biology curricula in order to meet the new standards."

Trouble in Kansas

As recently as November 2005, the State Board of Education in Kansas passed a similar measure. This time the vote was not unanimous, though the result was essentially the same: a mandate to teach all sides of what could now be considered a full-blown "controversy." Although in both cases there has been no curricular recommendation to teach a specific opposing view—such as Intelligent Design—the door that has been shut against Darwin's opponents for 80 years has, for the first time, been opened.

This followed a ruling in 1999 in which the Kansas Board eliminated almost all references to evolution in the curriculum—a move that Stephen Jay Gould likened to teaching "American history without Lincoln." Two years later, the board returned to curricular standards more sympathetic to evolution, before it recently reverted to its present position of the inclusion of Intelligent Design.

> **Evolutionary Revelations**
>
> Aside from its 1999 "evolution-elimination" ruling, the State Board of Education in Kansas moved to redefine science "so that it is no longer limited to the search for natural explanations of phenomena."
>
> —MSNBC, November 8, 2005

The national implications of this trend toward questioning evolution should be noted. Under the federal "No Child Left Behind Act," every state is required to conduct science evaluations every 5 years. Therefore, to the extent that there is sympathy for the Intelligent Design position, most states will have to face this ongoing controversy in their own backyards.

Problems in the Courthouse

In the Dover, Pennsylvania area school district, the issue went to court. The ACLU suit charged that the board was violating the establishment clause of the First Amendment, which prohibits the promotion of any religious doctrine. The board had required that students be informed about the theory of Intelligent Design, and in response, eight families sued to have Intelligent Design removed from the biology curriculum. Believing that the school board policy promoted a particular biblical view of creation, they contended that there was a violation of the constitutional guarantee of the separation of church and state.

The Discovery Institute was deeply involved in the case and came out against what it called "The ACLU's attempt to censor classroom discussion of Intelligent Design."

Invoking the landmark 1925 Scopes trial, Dr. John West, Associate Director of the Institute's Center for Science and Culture, released the following statement:

> Eighty years ago the ACLU went to court in Tennessee to defend the right of John Scopes to teach his students about evolution. Today, the ACLU is betraying the principal of academic freedom by seeking a government-imposed gag-order on teachers and students that would prevent even voluntary discussions of intelligent design in the classroom. All Americans who cherish free speech should reject the ACLU's effort to decide the debate over evolution through court orders rather than the free marketplace of ideas.

Nevertheless, U.S. District Judge John Jones ruled that the teaching of Intelligent Design would violate the First Amendment. "We have concluded that it is not (science), and moreover that ID cannot uncouple itself from creationist, and thus religious, antecedents," Jones said. He added, "To be sure, Darwin's theory of evolution is imperfect. However, the fact that a scientific theory cannot yet render an explanation on every point should not be used as a pretext to thrust an untestable alternative hypothesis grounded in religion into the science classroom …."

Man on the Street

Regardless of the way Intelligent Design is received in the schools and in the courts, the fact is that most of us on the street have our own opinions. If anything, it appears that the culture at large is more sympathetic to Intelligent Design than those in leadership positions in the public schools and higher education. Whereas teachers and college professors appear to resist such multiview approaches, the general population is more likely to advocate open academic discussion.

A Zogby International Poll taken in 2001 found that American adults heavily favored teaching the arguments against evolution. Only 15 percent thought that biology teachers should only teach evolution, whereas 71 percent believed biology teachers should also teach the evidence against it. Similar findings were met in Texas in 2003 and in California in 2004. A poll taken in Ohio in 2002, the year of the state school board debate, showed greater polarity between the two positions, with 19 percent of participants advocating that only Darwinian evolution be taught, and 65 percent standing in favor of both positions being aired.

This approach to education may well reflect the general population's take on evolution. In a November 2004 Gallup poll, only 13 percent of poll participants believed that God had no part in the evolution or creation of human life, whereas 38 percent said they thought that human beings evolved from less-advanced forms, with God guiding the process. Perhaps most startling is that fully 45 percent of participants contended that God created humans in their present form within the last 10,000 years.

Regardless of how one sees the debate playing out in the schools and in the courts, there is clearly "a fuss" that is growing more intense, and probably won't go away. What seems to be at stake is not only our cultural view of what science can and cannot explain, but our view of education, the place of religion in our society, and our understanding of democracy. Intelligent Design appears to be a lightning rod for American culture in this time; in order to engage in productive debate, we need to grasp the numerous complex issues surrounding ID.

The Least You Need to Know

- Creationism asserts that there is an Intelligent Creator, who is God; Intelligent Design asserts that there is an Intelligent Designer, though not necessarily God.

- Modern creationists, such as the late Dr. Henry Morris assert that science has erred, and that revised science may prove the biblical account of Genesis.

- The creationist movement is factionally diverse, and the various groups often disagree with one another regarding the particulars of creation.

- Creationism is a complete "theory," whereas Intelligent Design concedes there are questions yet to be answered.

- The ID debate in the schools, the courtrooms, and in the court of public opinion, will likely grow hotter.

Part 2

The History of ID

Despite what many of us may have believed about "the wall" between science and religion, the two have been—throughout most of history—intimately related.

In fact, most of the great scientists whom we hold in high esteem never excluded the idea of a "designer," that is, never until the advent of the science of the famous Darwin.

Chapter 4

Ancient Science

In This Chapter

- ◆ Eastern science
- ◆ Ancient Greek tradition
- ◆ The great Jewish "wheel"

Most of us living in the modern world assume that mathematics and science were engineered by the ancient Greeks—or at the very least—that these two interconnected disciplines are phenomena of Western civilization. That is true to a great degree, particularly when looking at "math" and "science" as we Westerners know it.

But there's more.

Dick Teresi's book, *Lost Discoveries,* which was discussed in chapter 3, explores the more ancient roots of Eastern science: Chinese, Indian, Egyptian, and Babylonian cultures. What is fascinating isn't simply the sophistication of these often disregarded cultures; it is the role that religion actively played in Eastern math and science.

"As for motivation, many scientific discoveries were driven by religion: Arab mathematicians improved algebra, in part, to help facilitate Islamic inheritance laws, and Vedic Indians solved square roots to build sacrificial altars of the proper size," writes Teresi. "This was science in the service of religion, but science nonetheless."

Science of the East

It is perhaps astronomy that is most naturally given to living side by side with religion: with its common focus on "the heavens" and its call to entertain the magnitude of the universe. In any case, religion played a starring role in Islamic astronomy, helping to develop ways of orienting worshipers toward Mecca. Sacred coordinates by which the shrine of Mecca was sighted represented a sacred use of math, and the shrine itself was positioned according to astronomical coordinates.

In the same way, the calculation of time was motivated by religion. Needing to compute the precise moments each day when worshipers were to pray, astronomy and math helped to order religious practice in Islamic culture. Elaborate tables to calculate these times superceded more empirical practices, which were eventually used with developing technologies—sun dials and compasses.

> ### Evolutionary Revelations
>
> In the Islamic world, the city of Mecca (in modern Saudi Arabia) is considered to be the holiest place in Islam. A pilgrimage, or journey, to Mecca is required of all "able-bodied" Muslims with the financial means at least once in their lives.

Cosmology

The study of *cosmology* has always been a religious enterprise. In its effort to describe the origins and history of the universe, cosmology has had to rely upon myths to explain what no one was around to explain. Teresi asserts that the ancient Eastern myths all resonate with scientific explanations; even the modern big bang theory was "primitively" rendered by Mesopotamian culture. (See Chapter 8 for more specifics on the big bang theory.)

Evidence that the study of the universe's origins—however scientifically expressed—is metaphysical by nature, may well be seen within the current big bang conflict.

def•i•ni•tion

> **Cosmology** is often said to be a metaphysical branch of astronomy that examines the cosmos, or rather the beginnings and the structure of the universe, as well as the "space-time relationship" within the universe.

Moving Beyond Science

Although "the big bang" is presently held as the most successful cosmological theory, cosmologists actively disagree about its metaphysical implications. Those who believe in God seem more content with the idea as being the only game in town, but as we shall see, those who do not believe in God are moved in very different directions.

Teresi points out that whereas cosmology is presumed to be "areligious," the nature of the discipline gives us little choice but to finally move beyond science. "We need to imagine our world, even if that vision is inaccurate or incomplete," he writes. "The ancient Indians, Babylonians, and Maya combined science and religion with social constructs to complete the picture. That we have done any differently is a delusion. If our cosmology appears free of religion, it's because we've made it into our own secular religion."

The Greek Greats

Because we are Westerners, the ancient culture that tends to get most of our attention is that of classical Greece. After all, to many minds ancient Greece is the unequalled civilization of Western history. Whether or not it truly is, its enormous influence continues to be felt in art, literature, government, and science, stemming from such philosophical luminaries as Socrates, Plato, and Aristotle.

It may seem that Greek philosophy is a far cry from mainstream science and Intelligent Design; however, we will discover that ID's way of "seeing" stems directly from this ancient tradition.

The various ways we modern people think resist the tendencies of these philosophers. But they resonate with thinking styles first developed by Plato, and differently by Aristotle. This is important not only because it clarifies from whence our ways of thinking come; it is important because Platonic and Aristotelian thought provide foundations for modern Intelligent Design.

It should be noted that not all ID advocates consciously adhere to such "Greek wisdom." However, it is clear that the assumption of design has roots in the fifth-century B.C.E., when Plato put forth his idea of "pure forms," and Aristotle, of "inductive reasoning." We shall see that these are schools of thought rather than ideas directly spoken from the lips of these two men (thus, the terms, *Platonists* and *Aristotelians*, rather than *Plato* and *Aristotle).*

Plato and the Neo-Platonists

Plato was born in Athens in 428 B.C.E., and died in 348. At age 40—following an education at the feet of his mentor, Socrates—Plato founded the Academy, from which derives the word *academic*. Located on a plot of land outside Athens, and owned by a Greek citizen, Academus, one of the earliest schools in Western civilization ran until the sixth century C.E.

It was here that Plato developed a system of thought that stressed analytical thinking, whereby truth was seen as embodied by "pure forms" that are abstracted from material existence.

Several centuries later, "neo-Platonism" expanded the realm of pure forms to incorporate a single, transcendent source from which the perfect, good, and true, must come.

Subsequent neo-Platonists held that physical matter bore an animating life, or soul. Objects were not simply dead, inert matter, but were imbued with active agents that explained such phenomena as motion, growth, and life in an otherwise mere physical existence. Thus scientific and religious beliefs were not separate or competitive with one another, but together illuminated the truth of the world. Science could be seen as the study of divine activity within nature, and the material world as the natural place wherein dwelt its "soul."

Fact or Faith _____

Copernicus may have been swayed by more than scientific calculation, affected as he was by the neo-Platonic spirit that inspired a "mystical" view of the sun.

Significant to the later Galilean debate about the sun-centered universe, neo-Platonists believed that the sun, in all its glory, had been placed at the cosmic center. The sun emanated light, even "enlightenment"—toward the truth of the Platonic pure form—and so this "heliocentric" perspective was only natural.

According to Nancy Pearcey and Charles Thaxton in *The Soul of Science*, "Copernicus wrote: 'In the middle of all sits the sun enthroned. In this most beautiful temple could we place this luminary in any better position from which he can illuminate the whole at once? He is rightly called the Lamp, the Mind, the Ruler of the Universe … So the sun sits upon a royal throne ruling his children the planets which circle round him.'"

In any case, neo-Platonism held that the realm of the metaphysical was inseparably connected to a physical world, which was sustained by this vital "mind," or "soul." As late as the seventeenth century, chemist Jean-Baptiste van Helmont understood carbon dioxide borne from burning substances as manifestations of God. Thus for many

Christians neo-Platonism became a way of perceiving God in the world, and explaining the relationship between the divine and material existence. "Christians in the neo-Platonic tradition tended to stress God's immanent presence in creation," write Pearcey and Thaxton. "Scientifically, that idea often translated into a notion of spiritual forces as the active power in natural processes. Gradually, those forces came to be seen as semiautonomous—implanted in creation by God at the beginning and thereafter capable of functioning on their own. Yet they still represented manifestations of divine power working in and through the created order."

Aristotle and the Aristotelians

Aristotle followed Socrates and Plato. In fact, Aristotle was Plato's student. Following Plato's death, Aristotle was allegedly passed over to head the Academy, eventually founding his own school, the Lyceum, in 335 B.C.E.

Aristotle differed from Plato both in style of thought as well as in the way he viewed science. If one characterized Plato's style of thinking as "deductive," that is—going from the general to the particular—Aristotle tended toward "induction"—moving from the particular to the general. Although Aristotle believed in abstract, formal truths much in the way of Plato, he refused to divorce the world of "pure forms" from that of matter, believing they were inseparable.

Unlike Plato, who distrusted sensory experience—that which was gained by touching and feeling—Aristotle liked to experiment, even practicing primitive dissection. Studying plants and animals and reflecting upon what he observed makes humans "human," Aristotle perceived the physical world as a whole, purposeful organism.

 Blinded By Science _____

Even prior to modern genetics, it was obvious that the development of living organisms must be directed by some internal pattern that ensures that acorns always grow into oaks, not maples, and that chicks always grow into hens, not horses. Based on this observation, Aristotle concluded that all forms of motion or change (all natural processes, as we would say) are directed by a built-in goal or purpose—a so-called final cause, which he also called an object's Form.

—Pearcey and Thaxton, *The Soul of Science*

God was no longer an unpredictable "soul" that vitalized inanimate objects, but a rational "mind" that could be apprehended by other rational, human minds. In contrast to neo-Platonic thought, Aristotelianism saw the earth at the natural center of the universe. "Everything in its place," seemed Aristotle's motto: so the natural place of fire was "up," whereas the natural tendency of rocks was to fall down to the ground to find their natural position.

In the Middle Ages, these Aristotelian forms became expressions of divine activity. However challenged they would be by the mechanistic perspective of the seventeenth century, we shall see that in modern-day "theistic evolution," Aristotle's influence is still felt. Aristotle championed the idea of "purpose" as a part of rigorous biology, making for a fruitful peace between science and religion—at least until the coming of Charles Darwin.

For the many differences that exist between the Platonic and Aristotelian philosophical schools, we can see that an obvious commonality is the presumption of a transcendent realm. Whether the realm is that of "pure forms" or of some singular "God," the material world did not preclude it; in fact, the material world depended upon such a realm of the transcendent in order to be understood. The Greeks saw active purpose in all creation, anticipating, by 2,500 years, the vision of Intelligent Design.

The History-Altering Hebrews

In his book *The Gifts of the Jews*, Thomas Cahill asserts that the Jewish perspective changed the Western world forever. The Hebrew perspective ran counter to most Eastern religions that surrounded these wandering people. So steeped are we in this vision of the world—of ourselves, and of what is possible—that most often we aren't aware of its profound influence, or that it could have been any other way.

The Jewish contribution to human history that Cahill convincingly conveys is that, by their wandering life, their experience of God, and the stories about themselves as people, the Jews departed radically from the Eastern world where they came to be as a community, opening up human history to an unprecedented understanding of progress. By virtue of their courage in the face of disaster, they glimpsed the possibility of a "future"—and came to perceive the "present" as full of potential for "newness." This sense of possibility not only constructed a fresh religious understanding, it enabled a dramatically new worldview that encompassed every aspect of human life and culture, even extending to science.

An Eternal Wheel

"The new" was an anomaly in the ancient world. It wasn't that "the old" was somehow better than "the new." These categories simply didn't exist as we now know them; according to Cahill. They were "newly" introduced by a developing early Judaism.

In the Eastern world where Judaism was born, life was understood as a wheel: moving in a never-ending cycle that returned to the same mythic life, again and again. Like other Eastern peoples, the Sumerians saw their lives as an "eternal return," and thus the point of living was to reenact the past in a way that would placate the mythic gods.

Science as a process of uncharted exploration, discovery, and unprecedented knowledge, would greatly benefit by the wandering Hebrew people for whom newness had become a way of life.

> **Evolutionary Revelations**
>
> Most of us in the modern world take it for granted that the "present," with its naturally implied possibilities for the future, is an intrinsic characteristic of all time. That seems to us an inherent given. But the idea that the universe, time, and human existence, was (or is) to be viewed in the sense of "newness," may be a distinctively Hebrew tradition.

The Judaic Here and Now

If for Eastern cultures, newness was a concept that was destined to upset their daily lives, by the Biblical accounts, it seemed to have become a cherished feature of the Hebrew people. The "present" was a landscape on which anything could happen, and anything might be discovered. So the future stretched as an ever-changing horizon, filled with hope for what was yet to be discovered.

Not only does science depend on the assumption of future discoveries; it equally depends on the assumption of a past from which new discoveries can be built. The open stance toward the future assumed by the Jews allowed for possibilities that scientists unconsciously embrace in the process of experimental discovery. Indeed, the notion of "scientific progress" presumes that there is such a thing as "progress"; in this regard, one might argue that we owe a debt to Israel for the history of science.

Though the Wheel Turns

In a sense, all cultures share a common need to take the material world seriously. People need to eat, and work, and sleep in shelter in order to survive. Yet what varies is the way cultures come to view this world in which they find themselves; and Judaism marked a departure from the views of their surrounding world.

The Eastern "eternal wheel," of which Cahill writes, turned freely above the material world and had little effect on the grit and accidents of unpredictable daily life. Rather than life being a succession of events that were unique and destined never to be repeated, it hinged on a fixed archetypal myth that was enacted regardless of the world "below." With the advent of a people who regarded daily life as a medium for human understanding, suddenly the world became an arena for all kinds of unprecedented human inquiry.

Though little science is recounted in the Biblical accounts that comprise the Hebrew scriptures, it is the Hebrew reverence for the earth as God's creation that made fertile ground for future science. Add to this the sense of uniqueness with which each character is purposefully rendered, and one finds a setting for unique human contributions that departs from a perpetual, unchanging past. Because the Hebrew story would be different if Moses hadn't died before he reached the Promised Land, all of human history, now largely driven by science, was opened to an uncertain future.

Opening of "Time"

Science, with its dependence on the material stuff of the world for its exploration, as well as its need to build upon successes of the past, has found fertile ground in twentieth-century American culture. Not only did the breaking of "the wheel" allow for the possibility of progress; it opened up a sense of time that Cahill argues encouraged modern scientific thought. The "big bang" theory, that says that the universe began at a singular moment, requires time to move in a linear fashion in order for the theory to make sense.

Indeed, the big bang theory is at the center of the Intelligent Design debate.

This Jewish perspective, grounded in "the new" and in a "future," has lent itself (perhaps by accident) to the fruitful practice of science. And rather than science and religion having to exist in separate realms, it could be argued that they come from the same place: a desire for human understanding.

"In a cyclical world, there are neither beginnings nor ends," writes Cahill in *The Gifts of the Jews*. "But for us, time had a beginning, whether it was the first words of God in the Book of Genesis, when 'in the beginning God created the heavens and Earth,' or the big bang of modern science, a concept that would not have been possible without the Jews."

The Least You Need to Know

- Many Westerners assume science and math were engineered by the ancient Greeks, but both disciplines have origins in the East.

- Astronomy in the Islamic world was developed to help worshippers orient themselves toward Mecca.

- Religion played a key role in Eastern math and science.

- Intelligent Design's approach to science is, in many respects, based upon ancient Greek philosophy.

- Plato believed that truth stemmed from "pure forms"—or "God"— and was abstracted from material existence; Aristotle believed "form" was inseparable from matter.

- The Hebrews came to see the present as having the potential for "newness," opening the way for our understanding of "progress."

Chapter 5

Medieval Science and the Great Masters

In This Chapter

◆ The great Aquinas

◆ The father of modern astronomy

◆ Newton's view

◆ How the masters saw things

We should understand that "theology" played a critical role in the practice of science until the period of the eighteenth century Enlightenment, and, in many ways it continues to play a crucial role in modern Intelligent Design.

Until the Middle Ages the church was predisposed to assume that knowledge came by faith in God, alone. However, this assumption, forged by Augustine almost a millennium before, was being challenged by the source of human reason. Convinced that reason and faith need not be in conflict but could complement one another, medieval philosopher and Saint, Thomas Aquinas, worked to philosophically and theologically reconcile divine revelation with rational thinking.

Arguably one of the most influential theologians in church history, Saint Thomas Aquinas (born in 1225 C.E.) came along at a fortuitous time when scholarly approaches to the universe began to demand rationally credible answers to questions about God.

He would ultimately help to shape fundamental philosophical arguments for the existence of God which are used to this day. For our purposes, he formulated what would become the foundational argument from modern Intelligent Design.

Right Man, Right Time

Aquinas believed that the existence of God could be proved. Based on Aristotle's idea of an "unmoved mover" and an "uncaused cause," Aquinas elaborated proofs of the existence of God in five discrete directions. Aristotle had said that there had to be a "first" movement which set physical bodies in motion—a first falling domino which caused every next domino to fall in the great causal chain of events which leads to this day.

From this idea came Aquinas's arguments for God.

The arguments were made as follows:

- ◆ The argument from motion
- ◆ The argument from efficient causes
- ◆ The argument from possibility and necessity
- ◆ The argument from gradation of being
- ◆ The argument from design

Recognize a word in the last argument? It is this last argument—the medieval, philosophical argument from design—that we see used by proponents of Intelligent Design more than 700 years later.

Toward Purpose, Not Chance

The first assumption of the philosophical argument from design is that the material stuff of the world—specifically, "natural bodies"—interacts in the world with some apparent purpose.

Evolutionary Revelations
Thomas Aquinas often used the analogy of an arrow fired from a bow that reached its destined target, indicating there had to have been an archer who took aim and released the bowstring and the arrow, thereby guiding the arrow to its purpose.

To Aquinas the universe functions like an arrow shot toward a target, and is sustained not by random accident but by some manifest purpose that is evident in the order of the world we all see.

In philosophy we call this a *teleological* argument. So using this idea, Aquinas asserted that the world seemed to work toward some purpose.

def•i•ni•tion

Teleology derives from the Greek word *telos*—which literally means "end" or "purpose."

Most Things Lack Knowledge

The next step in Aquinas's argument was to establish that natural bodies lacked smarts. Arrows are inert—just as dumb as they look—lying there on the ground. If they were left to their own devices, they would go on lying there until the end of time.

It should be noted that for Aquinas, human beings were somewhat different. Human bodies were vessels of the soul, reflecting God's activity. In any event, it was clear to him that matter by itself lacks its own volition; something else was needed to finally explain its apparent purpose which we see in the world.

Intelligent Direction?

Aquinas said if an arrow reached its target, it could only do so by design. Something else had to have set the arrow in motion and guided it to its intended destination. In other words, matter (the deadwood) is the stuff that is finally infused with or affected by the intentional design of its creator. It (the arrow) reaches its goal (the target) by being directed by some agent of intelligence (the archer).

In scientific terms, Aquinas might have said that biological life shows apparent design indicating a designer who must have brought it into being.

Intelligently Inspired

Because the physical world exhibits this design, and appears always to move toward some purpose, Aquinas asserted that there had to be an intelligent being that inspired it. Sounds reasonable?

Be it an arrow flying toward a bull's eye, or the complex magnificence of the physical world, Aquinas saw proof of an intelligence that clearly lay beyond both. And for Aquinas (more than 700 years ago) this intelligence was God, the great Intelligent

Designer who could be seen in every facet of the physical world.

This medieval argument serves as a foundation of Intelligent Design.

So there is really nothing new about the contention that the physical world reflects a "God." What is new is the scientific world in which this argument suddenly finds itself. For some, contemporary research is unprecedented proof that there is a God behind it all. For others, such research provides final proof that God simply does not exist.

The Great Galileo

Prior to the sixteenth century—when Polish astronomer Nicolas Copernicus suggested that the sun was at the center of the universe—the earth was considered by all to be at the universe's center.

Copernicus, of course, upended what was considered to be a "geocentric cosmos" (which was inherited from the Greek philosopher, Aristotle, whose position was that "it's all about us"). But because this was only a theory that had yet to be proved, the world would have to wait until the following century before someone would come along to vindicate his calculations.

Enter the great Italian physicist, philosopher, and astronomical genius, Galileo Galilei.

The Heliocentric Universe

Born in 1564 C.E. in Pisa (yes, the home of the famous leaning tower), Galileo has since become known to us as the father of "modern astronomy," "modern physics," and "modern science." As such, he has become a central figure in the modern debate over Intelligent Design.

With the use of his primitive telescope and his gift for geometry, Galileo revealed to the world that our cosmos was "heliocentric": that is, the sun was indeed (as Copernicus had suggested) at the center of the known universe.

By the power of mathematics, his ingenuity, and his empirical investigation, Galileo inserted this revolutionary truth into his very Roman Catholic world.

What About the Scandals?

Ah, yes, the first thing that usually crosses the minds of scholars when they hear mention of Galileo's name is his conflict with the church. Yet despite what many of us may have heard regarding this conflict and his "scandalous beliefs," many such accounts are overblown, or simply wrong.

The Indigo Girls' song "Galileo" notwithstanding, Galileo never left the church, and as much as the church may be resented these days for Galileo's arrest, it is important to keep the facts straight. Not that the church didn't act like every institution in its need to exercise control, but it wasn't alone in its resistance to change and its need for self-preservation.

It is true that Galileo was placed under house arrest for his "scandalous" sun-centered cosmos. But despite the many paintings in which he is depicted being brought before intimidating hoards of clergy, he actually appeared before a tribunal of two officials and a recording secretary. Rather than being put in chains in a dungeon, as has also been frequently portrayed, Galileo was in fact "sentenced" to the Tuscan embassy in Rome, and then to the archbishop's palace in Siena. There he worked for 25 years on a book he had wanted to write, living in relative comfort, and often in the presence of other scientific scholars.

Evolutionary Revelations

… instead of holding Galileo prisoner as a confessed heretic, (the Archbishop) indulged him as a guest of honor. "The Archbishop," an anonymous hand informed officials in Rome, "has told many that Galileo was unjustly sentenced by his Holy Congregation, that he is the first man in the world, that he will live forever in his writings, even if they are prohibited, and that he is followed by all the best modern minds."

—*Galileo's Daughter: A Historical Memoir of Science, Faith, and Love* by Dava Sobel

---

Galileo Refuses to Recant

Galileo was devout in his Catholic faith to his death and buried in a Franciscan church. The view that Galileo was opposed to all church teachings is not true. Like many others, he was opposed to some church teachings—but in particular, the one that claimed a geocentric cosmos.

Of course we always look at history from the present, and see it through contemporary lenses. The lens that we've been given is one that was ground by the conflict between science and religion, and so it is only natural that the Galilean affair looks the way it does. The white knight of "science" has been pitted against the dark knight of "the church," and Galileo's faith has therefore been revised to suit our own contemporary cultural belief in science.

We might view Galileo's Christian faith as a quaint artifact of the time that is irrelevant to his science. But it's important to ask whether this is an honest move: taking "the good science" without "the bad religion." We should at least ask whether, to be fair, we must allow a bigger picture in which Galileo's science looks pretty "religious." Whether we like it or not, this master of science believed in God to his death, and (one could argue) he perceived his faith as a critical component of his work.

Galileo's War with Secularists

Galileo was living in a world governed by Aristotle. As many Galilean supporters as there were in the Catholic Church at the time, secular Aristotelian philosophy, which placed the earth at the center of the cosmos, had critically influenced the faith. Moreover, it is important to note that this geocentric universe was based upon scientific understanding (however wrong it now seems) as well as upon religious faith. In this regard, this was as much a scientific as a religious debate—and in fact, history further tells of distorting academic politics reminiscent of our own day.

> ### Evolutionary Revelations
>
> … a lively struggle was taking place between an older elite in the universities and churches, and the newer, more pragmatically oriented elite to which Galileo belonged. Galileo's decision to publish his works in the vernacular (rather than in Latin) was a deliberate affront to the established elites, part of a broad strategy to transfer intellectual leadership to the wider reading public.
>
> —*The Soul of Science,* by Nancy R. Pearcey and Charles B. Thaxton

God—Not Man—at the Center

For others, Galileo's sun-centered cosmos simply was not a problem. Neo-Platonists held that the sun was at the universe's center. In fact, Copernicus, the first to argue for a heliocentric cosmos, was convinced not by scientific evidence (of which there was none) but by his commitment to neo-Platonism.

Fact or Faith

In the middle of all sits the sun enthroned. In this most beautiful temple could we place this luminary in any better position from which he can illuminate the whole at once? He is rightly called the Lamp, the Mind, the Ruler of the Universe ... So the sun sits upon a royal throne ruling his children the planets which circle around him.

—Copernicus

Science writer Dick Teresi notes that the earth was not a "favored place," but the toilet drain into which the swill of the universe was emptied. It was the sun, distant in the heavens and emanating warmth and light to all creation, that made sense to followers of neo-Platonism as it expressed itself in the Christian faith. In contrast to the belief that the earth-centered universe was a sign of the church's arrogance, here man in all his sinfulness was at last where he belonged.

Galileo's God as "Divine Architect"

Rather than seeing God as a problem that got in the way of science, Galileo saw God as the "divine architect" who was responsible for the universe. Never is there doubt in Galileo's writings that God existed, nor did God disable "good science." To the extent that there was conflict, it was with the institution of the church, and particularly church authorities. In the scientific time in which we live it is difficult to understand how this was so, and even more difficult to understand how science could have accommodated God.

Just as he was a scientist to the end, Galileo was a Christian. The physical universe was God's domain that could not be questioned by science; to Galileo, the scientific truths which he discovered himself, were testimony to its architect. It may be fair to conjecture that, at the very least, he would be confused by the modern conflict between science and religion—in a world that he believed was living proof of God.

Sir Isaac's Science

There was also, of course, Sir Isaac Newton, whom you have to blame for most every boring physics lab you ever had to take. Dropping balls from lab tables and rolling them down the halls is the fault of Newton—the late-seventeenth and early-eighteenth-century English scientist. Although most of Newton's laws seem self-evident now, they weren't when he was in business.

At the time Sir Isaac lived, it wasn't clear that his mechanical laws could even be formulated. Whereas now they seem as natural as breathing, or as familiar as the kitchen sink, their discovery marked a dramatic new chapter in science in which the world was seen as universally predictable.

Blinded By Science

In Newton's view, the world resembled a pool table filled with billiard balls that moved and interacted by mechanical principles, rather than by shoving from within. Gravity for Newton would no longer be seen as an interior feature of the object but as a physical law that acted from without—upon billiard balls, labs, and physics students.

Like Galileo, Newton was also living in a world governed by Aristotle, whose followers were given to believe that objects had their own inherent structures that determined how they acted in the world. They thought that motion was a function of an object's "natural tendencies," rather than resulting from external natural laws that controlled how they moved from outside them.

The universal power of Newtonian mechanics to predict the way that objects would behave outdid Aristotle's unfounded abstractions about the ways objects moved in the world.

Newton Believed in God

Newton could easily have been presumed to be an unreligious man. With his mechanistic laws successfully predicting the motion of any worldly object, should this not have banished the need for anything supernatural—for any God at all?

Shouldn't the mechanistic science of Newton—who finally uncovered the natural physical laws—have allowed objects to act in the world "on their own" without the need of an urge from a God?

The answer is no. In fact, Newton presumed God's existence much in the same way that Galileo did. As we shall see, he even opposed the "distant God" who was promoted by the deists of the time, those who believed that God wasn't an ongoing player in creation.

God as the "Great Engineer"

In *The Soul of Science*, authors Nancy Pearcey and Charles Thaxton explain that Newton perceived God as the great engineer, "very well skilled in mechanics and geometry." This isn't like anything most of us learned in high school or college physics classes—perhaps because science in our day has become so secular an enterprise.

Newton himself was extremely religious, using his research to prove the existence of God, rather than God as an illusion. Like all scientists, he was doing science in a culture of competing philosophical perspectives. The neo-Platonists and the "mechanistic" thinkers were shooting from both sides of the bow, and in the end Newton found a way to embrace the virtues of each point of view.

Squeezed by Plato and Aristotle

The active principles in which the neo-Platonists believed were a sign of God's participation in the world, and so the world was always dependent upon God in a fluid, intimate way. A good example of this comes from Jean-Baptiste van Helmont, a seventeenth-century chemist, who thought all objects had souls that survived their own destruction in the form of some kind of gas. In this way he contradicted scientific materialism, asserting a spirituality that transcended the hard boundaries of concrete objects to live on in the realm of God.

On the other side of the bow, thoroughgoing mechanists took fire by the opposite approach. Matter was benign, having no creative power, with objects interacting with each other; the spirit world was separate, and lived unto itself in an untouchable domain of the divine. As a result, Renè Descartes took a mechanistic view of the world, rejecting miracles as dirty worldly work that God would never stoop to have a part in.

> ### Evolutionary Revelations
>
> Renè Descartes was a French scientist, mathematician, and philosopher. Also known as Cartesius, Descartes developed the Cartesian system used in geometry and algebra. Descartes is the "founder of modern philosophy" and the "father of modern mathematics."

Sir Isaac put together these two different approaches with his fellow "Cambridge Platonists." As the guru of mechanical physics who is credited with showing how the world could be godless—from tidal calculations to mathematical computations of planetary motion—he was clearly a believer in the natural laws governing the world in which he lived, but resisted the idea that it happened without God.

Always searching for divine activity, Newton found it in what he called, "supramechanical or active principles in nature … [that were] revealed in such phenomena as chemical reactions, light, and magnetism." (*The Soul of Science*, by Nancy R. Pearcey and Charles B. Thaxton)

But gravity, the force for which Newton is most famous, reveals best how he joined the two approaches of neo-Platonism and a mechanistic worldview that spoke together of the activity of God. Gravity was a mechanical force that was governed by the laws that Newton discovered. But equally, it showed an active dynamism that finally revealed an active God: utterly invisible, and so, mysterious—yet predictable as any engineer.

Not a Deist or Rationalist

Deism challenged the Christian church, asserting that natural law and reason were the way to an afterlife. The path that the church had constantly asserted was that of dogma and doctrine, which as we all know, held to the performance of miracles by Jesus. But deism threw out the literal truths of the Bible, arguing instead for a faith based upon empirical evidence—the natural machinations of the world. Deism held that God didn't have such an intimate relationship to the world, relying instead upon a distant deity that simply set the worldly laws in motion.

Newton took pains to avoid the deism that was prevalent in his time. He believed that God indeed participated in the ongoing affairs of men. He rejected the idea that the world was "mechanistic" in the way of deistic thinkers, because he felt it would have diminished the power of the divine to govern human beings.

Nor was Newton a rationalist. Rationalism—raised to an art by Descartes's "I think therefore I am"—rejected the material world as a reliable source of human knowledge. Only abstract logic, best expressed by mathematics, was a worthy means of understanding.

This set the foundation by which the world could be understood. Newton believed that God created such logic, imbuing human beings with this perfect capacity in order to meet the perfect nature of "himself."

Reason over Superstition

The Enlightenment, or "the age of reason," in which Newton lived his adult life, championed reason over the "superstition" still emanating from the Middle Ages. It put man and his unique capacity for reason front and center on the stage of life, appealing to his flourishing intellectual powers which Newton himself exemplified. Still, Newton held fast to his belief in a God as actively engaged in the world—despite his own unprecedented work to describe a mechanical universe.

Blinded By Science _____

> [Newton writes] that the business of science is to "deduce causes from effects, till we come to the very first cause, which certainly is not mechanical." In Newton's eyes, the major benefit of science was religious and moral. It shows us "what is the first cause, what power he has over us, and what benefits we receive from him," so that "our duty towards him, as well as that towards one another will appear to us by the light of nature."
>
> —*The Soul of Science,* by Nancy R. Pearcey and Charles B. Thaxton

The Great Masters' Perceptions

We've learned the history of science is full of different ways of going about scientific work. Galileo stands as a critical beginning of the way that modern science has been done: using logic, calculation, measurement, and ingenuity by way of telescope, his scientific method proved Copernicus's beliefs and a heliocentric universe.

Newton found the mechanical laws that describe how objects move in the world, by way of a scientific method that is still used in physics classrooms today.

Galileo Committed to God

As noted, Galileo never left the church, despite the way it treated him. Rather than blind obedience, we see in him courageous admiration: for the world in all its majesty, and its natural laws, which he believed could only come from God. His science found no conflict with his religious faith; in fact, it was an integral part.

Newton's Writings

Newton's writings brimmed with theological allusions, as though they were the very air he breathed. His fight was not between science and religion, but between competing religious perspectives. So he would reject deism as a danger to the sovereignty of his God.

Science Spun by Those to Come

Fact is, the perceptions of the great masters of science have been clouded by those who came later.

In Newton's case, later Enlightenment thinkers spun him in secular directions that better fit the mechanistic worldview of the day—which, three hundred years hence, is still with us.

In terms of Galileo, we ourselves may be principally responsible for the way in which he is remembered: we see him as a victim of injustice who defied the church, yet prevailed by the sword of science.

Both men, in fact, blended scientific processes with their religious faith, perhaps paving the way, 500 years later, for modern belief in "design."

Fact or Faith _____

In the cosmic order, Newton saw evidence of intelligent design. "The main business" of science, he said, is to argue backward along the chain of mechanical causes and effects "till we come to the very first cause, which certainly is not mechanical." Newton also regarded several specific characteristics of the world as inexplicable except as the work of a Creator. "Was the eye contrived without skill in optics," he asked, "or the ear without knowledge of sounds?"

—*The Soul of Science,* by Nancy R. Pearcey and Charles B. Thaxton

The Least You Need to Know

♦ Thomas Aquinas believed that the existence of God could in fact be proved.

♦ Nicolas Copernicus suggested that the sun was at the center of the universe. Galileo Galilei proved it.

♦ The widely held view that Galileo was opposed to all church teachings is not true. A scientist of the highest order, he was also a devout Christian throughout his life.

♦ Galileo saw God as the "divine architect" of the universe.

♦ Sir Isaac Newton saw God as the "divine engineer" of the universe.

♦ Intelligent Design is anticipated by the work of Galileo and Newton.

Science in the Age of Reason

In This Chapter

We've all heard of the time in history called "the Enlightenment" or "the Enlightenment period" or "the age of enlightenment." We've also heard of the "age of reason."

Enlightenment? Reason? The two periods sound related don't they? It's because they are. Though some historians consider the age of reason a sub-part of the overall Enlightenment Age, they are more widely considered to be the same period, with different names.

So what was the Enlightenment?

It was just as it sounds: a time in European history (and to a lesser degree, colonial American history, because the first great awakening was about to burst forth in the colonies) when belief in the power of rational thought, or reason, reigned supreme.

Reasonable Science

Following the Middle Ages, when the church dominated daily life—from commerce, to art, to intellectual thought, to the boundaries of religious belief—the Enlightenment reacted against the church's power with a liberating secularism that elevated human reason above church authority.

Science as a facet of the Enlightenment belief in the possibility of human progress marked the intellectual temperament of the seventeenth and eighteenth centuries. As a developing engine for human invention and a burgeoning source of knowledge, science would receive increasing attention as a medium for human understanding. In an effort to define scientific inquiry as a rigorous discipline unto itself, English philosopher Francis Bacon published *Novum Organum* in 1620.

The Need for a "Scientific Method"

As the politician that he was, Bacon believed in practical kinds of knowledge. Reacting against empty philosophical speculation as a sign of human weakness, he rejected theoretical scholasticism in favor of pragmatic understanding. Emerging from a time when abstract theological dictates often went unquestioned, the intellectual freedom of the Enlightenment was fertile ground for the reign of scientific proof.

As an amateur scientist as well as a philosopher, Bacon believed in the empirical, shunning generalized abstractions in favor of concrete experience. Identifying "idols" that impeded an effective understanding of the world, he argued for objective truth over and against the "superstitions" of theology and myth. In this spirit, Bacon set out to develop a method which transcended such impediments; and so, the Baconian "scientific method" would be ultimately formalized.

> **Evolutionary Revelations**
>
> Francis Bacon coined the phrase *knowledge is power.*

Inductive Reasoning, Empirical Investigation

The foundation of Bacon's scientific method was a dynamic combination of empirical knowledge—knowledge from experience—and inductive reasoning. What is significant is that Bacon elaborated rules for scientific experimentation, by which truth could be apprehended beyond the deceptions of subjective prejudice. Until this point, no such formal methodology had been constructed to so define science. Now there was a standard for scientific practice about which scientists could agree.

Empirical knowledge is knowledge that derives from observation of physical phenomena. Unlike rational thought, empirical knowledge is gained by material that can be observed, and so science was a process of collecting data from empirical observation. By such observation and experimentation, one built up, through inductive reasoning, principles about the physical world that would hold in other situations.

Inductive reasoning (the other half of Bacon's methodology) is the opposite move to thinking by deduction, against which Bacon was reacting. Unlike most theologians, who begin with universal abstractions—God is good, therefore I will go to heaven (assuming I am good)—Bacon started with particular, concrete, physical phenomena, from which he would extrapolate general principles that revealed themselves as universally true.

Testing Science

One of the critical arguments repeatedly raised in the Intelligent Design debate has to do with characteristic differences between science and religion as distinct modes of human inquiry.

From mainstream scientists we hear, "religion isn't testable." Scientific inquiry, scientists claim, requires that hypotheses be tested. Religious faith is scientifically meaningless because it isn't testable; so the proposition of an "Intelligent Designer" is inherently untestable.

Indeed, proponents of Intelligent Design agree that scientific propositions need to be testable. Their disagreement hinges on whether or not Intelligent Design is testable.

> **Evolutionary Revelations**
>
> Empirical "testability" is a defining feature in the modern debate about what precisely constitutes science.

Paley's Watchmaker Universe

The Enlightenment naturally grew out of the enormous success of Newtonian mechanics. The world was suddenly understandable by seemingly miraculous laws no longer requiring Aristotle's mysterious ideas of how the world must work. There had been other philosophical whims in the air, such as those of the neo-Platonists, who believed in active principles, or "souls," that had their own vitality. Now there were signs that the world had an order that was free of a God in its backyard. And even if there was a God, the world was created so that it could pretty much run on its own.

def•i•ni•tion

Transcendent comes from the word *transcend*, which means to rise above or go beyond the universe, time, or a specific dimension.

English philosopher William Paley published *Natural Theology* in 1802. He proposed that the world was like a giant watch that was governed by these natural laws. Typical of much of the thinking during the Enlightenment, Paley expressed the common belief that whether or not one believed in a God, the universe clearly possessed a *transcendent* order. As we shall see, Paley's watchmaker universe was a transitional idea—from serving as proof of God's existence, to being the end of any need for a God participating in the world.

God as a Watchmaker

The watchmaker argument is in many ways the foremost icon of the Enlightenment. It is mentioned here because it is often used in arguments for Intelligent Design. It isn't science, but a philosophical analogy that illuminates the age of reason—in all its power, and its limitations as revealed by modern science.

The watchmaker argument is an attempt to describe the world in which we live. Whether intended or not, it reveals the Enlightenment assumptions of the eighteenth century, and sets the stage for the coming battle between science and religion. Assuming that the universe is like a watch, its inherent ingenuity "naturally" implies the presence of a God.

The Watchmaker's Universe

"In crossing a heath, suppose I pitched my foot against a stone, and were asked how the stone came to be there; I might possibly answer that, for anything I knew to the contrary, it had lain there forever: nor would it perhaps be very easy to show the absurdity of this answer. But suppose I had found a watch upon the ground, and it should be inquired how the watch happened to be in that place; I should hardly think of the answer I had before given, that for anything I knew, the watch might have always been there."

This passage by William Paley strived to show how the natural world indicated the existence of God. For those who held a Christian belief in God, this was vindication of a creator. And for deists (as well as for atheists), it was an intimation of natural laws that could provide a lens through which to look at the world, in its ingenious design.

So Paley would go on: "Every indication of contrivance, every manifestation of design, which existed in the watch, exists in the world of nature; with the difference, on the side of nature, of being greater or more, and that in a degree which exceeds all computation."

Fact or Faith _____

Paley's analogy is presumed by a majority of Americans most of the time in their lives. In a 1991 Gallup poll, 90 percent of Americans interviewed believed that God created the world. While undoubtedly some learned that this was "true" in Sunday school, still others saw signs of design in the most ordinary, everyday things, which ultimately lead to a creator.

Nature Designed Like a Watch?

From the eye of a frog, to a snowflake, to the moon in its gravity hanging in the sky, we're unconsciously inclined to believe that something "other" created the world. Looking at the biological order, in its many miraculous expressions, gave Paley to liken it to an object that we all know has been designed. He went on to explain that life's organisms surpass a watch's intricacies, and so we are left to deduce that there has to have been an intelligent designer.

Complex Objects Created

Paley was already contending with those who believed only in physical causes. English philosopher David Hume, preceding Paley by several decades, espoused the belief that knowledge is no more than a series of sense impressions. And dynamics such as *causality* are only artificial mechanisms we make up to cohere the world around us.

Paley takes a different tack from Hume: in which causality's the name of the game. Exploiting the intuition of the man on the street (that causality simply "works"), he argues our most natural assumption is that there is design in the world. He points to the absurdity of coming on the intricate watch lying on the ground and believing it must be the result of natural forces—specifically, of wind and erosion. If one deduces that a complex watch was created by human intelligence, he asserts, then the complex larger world must be seen as a consequence of divine intelligence.

def•i•ni•tion _____

Causality comes from the word *cause*, which means a person or thing acts, happens, or exists in such a way that something specific happens as a result. For instance, if an Intelligent Designer acted on something that resulted in the big bang, the Intelligent Designer's actions would be that which "caused" the big bang to take place.

Science: What Is It Now?

Science is science: right? It's detached; it's objective; it's true. Science is simply what you do to finally get "the facts."

Yes and no.

Yes. Science uses certain techniques to make sure it's getting it right. In high school classes, most of us learned the basic techniques of science—observing the world, making a hypothesis, testing it, and revising—so that at the end of a day in the lab, we've gotten a little bit closer to understanding the facts, the objective truth, and the way that the world really works. What could be more unbiased than the objective way of science; what could be less inclined to allow our subjective prejudices?

And …

No. If there were such a thing as being "objective" in any ultimate sense, there probably would have been an objective definition of *science* long ago. Yet returning to Dick Teresi's quest for such a definition, there doesn't seem to be. "[The American Association for the Advancement of Science], for example, does not have one. After many trials, the American Physical Society [for physicists] finally decided upon a definition. The APS found that if the definition was too broad, pseudosciences like astrology could sneak in; too tight, and things such as 'string theory,' evolutionary biology, and even astronomy could be excluded."

def•i•ni•tion

Science is:

1: the state of knowing: knowledge as distinguished from ignorance or misunderstanding.

2: a: a department of systematized knowledge as an object of study. b: something that may be studied or learned like systematized knowledge.

3: a: knowledge or a system of knowledge covering general truths or the operation of general laws especially as obtained and tested through scientific method. b: such knowledge or such a system of knowledge concerned with the physical world and its phenomena.

4: a system or method reconciling practical ends with scientific laws.

—*Merriam-Webster, 2006*

Given centuries of doing science, one would think that someone would have come up with a definition upon which scientists agreed. Yet the problem may lie not in the elusive nature of science itself but in our inevitable human "subjectivity": that each of us stands in one place. As rigorous as a lab may be in testing how the world works—apart from our wishes, our preconceptions, and what we ate for breakfast—still we have no choice but to bring our own particular eyes to the lab, and to draw conclusions that make sense to us in the context of our uniqueness.

The General Scientific Assumption

Of course part of successfully doing science is achieving scientific agreement. Until a theory is shown to be true for other scientists in their own labs, it isn't "successful"—which is to say, it isn't accepted as "true." Truth depends upon its being shared by a community of scientists (and perhaps even us, who tacitly agree about things such as gravity every day).

A truth that is shared by most scientists these days is that God isn't part of the equation. According to a 1998 survey published in the journal *Nature*, only 7 percent of the responding scientists believed in the existence of God. This contemporary survey replicated a landmark survey of 1914 when 28 percent of scientists held to a belief in the transcendent. A survey taken in 1933 predictably shows a middling result: 15 percent of scientists surveyed believed that there was a God.

What is interesting is the conclusion drawn by the authors of the most recent study. They conclude, "As we compiled our findings, the [*National Academy of Sciences*] issued a booklet encouraging the teaching of evolution in public schools, an ongoing source of friction between the scientific community and some conservative Christians in the United States. The booklet assures readers, 'Whether God exists or not is a question about which science is neutral.' NAS president Bruce Alberts said: 'There are many very outstanding members of this academy who are very religious people, people who believe in evolution, many of them biologists.' Our survey suggests otherwise."

def•i•ni•tion

The **National Academy of Sciences** (NAS) is an honors society of distinguished science and engineering scholars involved in both scientific research and NAS's commitment to further the practical uses of science.

Consciously or not, the authors of the NAS study make the significant point that modern science isn't "neutral" about God—that it has a clear theological position.

The position seems to be that God doesn't make sense. Though some scientists may say their lack of faith has nothing to do with their work, it seems most are inclined to assert that their science has inspired their atheism.

If this is the case, it reveals an assumption that science can answer "theological" questions. Although twentieth-century scientific answers about God tend to be made in the negative , science still stands in the theistic tradition of Galileo and Newton. Perhaps this is why science so often brushes up against God: it has so long been a critical mode of theological inquiry.

No Room for God

This was the world into which Charles Darwin was born. If its romantic biology had sprung from a radical rejection of mechanistic science, its burgeoning mechanistic biology was equally induced by romantic, even pantheistic science. The unproven "life force" of the neo-Platonists, which seemed a vague sort of "God of the gaps," began to give way to a resurgent need for mechanistic certainty.

And Darwin met this need. As we shall see in greater detail (see Chapter 14) when we look specifically at Darwin's theory, random mutation and natural selection silenced the need for God.

Blinded By Science

When a "divine plan" in his evolutionary theory was suggested by a fellow botanist, Darwin objected, saying that this was not what he meant to say. Darwin expounded the importance of mutation being random, and so selection being based upon chance. "If the right variations occurred, and no others, natural selection would be superfluous."

The brilliance of Darwin's evolutionary theory was that it revealed how life could run on its own. If the godly watchmaker was increasingly removed from intervening in human affairs, there was still the recognition that the watchmaker was there—to remind us of whose watch it was. But if natural selection were a self-sufficient machine, biology could run without God: explaining who we were, where we came from, and where (without God) we wouldn't be going.

Therefore, part of Darwin's project was to get God out of the science. Apparent design that intimated a "creator" was seen as an illusion, to be replaced by natural "adaptation" which could define an organism without Intelligent Design. Aristotle's idea of a "perfect fit" between an organism and its environment was superceded by "survival of the fittest" characteristics for a particular environment. In the same way, the neo-Platonic "archetypes" put forth by the romantic biologists—in which universal "forms" would direct the evolutionary process to a destined end—were supplanted by the idea of a common ancestor that could explain our material beginnings.

Redefining "Science"

Darwin not only redefined the way organic life had been constructed; he redefined the medium—the way that science worked—and so the face of biology. He didn't simply vanquish competing ideas, such as Plato's ideal "forms" or a Christian "God"; by asserting that science had no room for such abstractions, his idea out-survived the godly competition. An empirical, natural world which was governed by mechanistic forces was the only world where science could now operate, and so the only world that was real.

The struggle to find purpose, or teleology, was resolved "once and for all." Design was illusion, or in the words of later neo-Darwinists, just "apparent." Accommodating God was a waste of time that "real" science could no longer tolerate; so, departing centuries of theistic tradition, Darwin changed the face of science itself.

> ### Evolutionary Revelations
>
> This denial of purpose is Darwin's distinctive contention ... The sum total of the accidents of life acting upon the sum total of the accidents of variation thus provided a completely mechanical and material system by which to account for the changes in living forms. To advance natural selection as the means of evolution meant that purely physical forces, brute struggle among brutes, could account for the present forms and powers of living beings. Matter and force ... explained our whole past history and presumably would shape our future.
>
> —Philosopher Jacques Barzun

The Evolution of Social Science

One of the effects of Darwinian thought that we don't like to talk about is how it was applied to social relationships and societal constructs. Here strange bedfellows in the evolutionary story were liberal academic evolutionists who suddenly found themselves in the sack with nineteenth-century robber barons. Indeed, the sinister story of social Darwinism still chases modern-day academics, who continue to hold to Darwin's "survival of the fittest" with its cruel and often racist implications.

Herbert Spencer, an English philosopher (1820–1903 C.E.), convincingly applied Darwin's theory to human development. A member of the ruling class, Spencer saw the presence of worldly wealth and power as the fruits of natural selection. The competition implied by "survival of the fittest," Spencer went on to explain, benefited mankind by removing from the human pool the weak, and therefore the "unfit."

Throughout history, he said, society operated like a jungle in which the fittest survived. Yet as cruel as this might be, it would gradually evolve us toward a just and peaceful existence. For the fittest, in Spencer's evolutionary view, were those most capable of living in peace; so evolutionary theory was beginning to be used to describe utopian visions.

Antisocial Darwinism

With the rise of unprecedented capitalists such as Andrew Carnegie came a new use for Darwin's grand and all-embracing theory. Within the morally sketchy economic arena of unbridled competition, "survival of the fittest" was a welcome scientific justification for its evident greed. Unrestrained competition was no more and no less than natural selection at work, progressively improving the socioeconomic system by weeding out the unfit.

> **Evolutionary Revelations**
>
> Between 1870 and 1900, we see the height of *laissez-faire* capitalism, undoubtedly reinforced by the moral vision implied by "survival of the fittest." Although untold wealth was made, it was being made on the backs of impoverished laborers who worked 80 hours a week, and commonly got sick and died from the conditions of the workplace. Social Darwinism, convincing as its original science may have been, had begun to show its colors and experience its manifest social implications.

Thanks to a general social outcry and the rise of labor unions, laissez-faire capitalism began to wane at the close of the century. But this was not before the Supreme Court had repeatedly backed the wealthy capitalists, ruling in favor of unconstitutional state laws regulating maximum work weeks, minimum wage, and child labor. There were even admonitions not to give to the poor so as not to encourage "laziness," thereby spawning a work ethic motivated not by God, but by one's "biological" nature.

Therefore, Herbert Spencer would proclaim: "… the strongest and the fittest should survive, while the weak and unfit should be allowed to die." In such a spirit, to help the weak was deemed immoral because it functioned to promote the survival and reproduction of someone who was unfit. Spencer asserted that society advanced when its fittest members were allowed to "assert their fitness with the least hindrance," while the unfit "should not be prevented from dying out." In fact, Spencer concluded that helping the poor needlessly delayed their extinction.

Social Darwinism in the Twentieth Century

While Social Darwinism was flagging in Europe, it found new life in other places. As Britain was practicing its imperialism in countries around the world, Darwinism proved a significant justification for conquering weaker, or "less-fit," peoples. Inevitably, racial discrimination became "scientifically" rationalized in the minds of many Darwinian adherents, ironically undermining the Darwinian promise that natural selection would lead to world peace.

Perhaps the bleakest outcome of Social Darwinism was the discipline of eugenics. Operating from the assumption that certain racial and ethnic groups are genetically "superior," eugenicists believed they could "improve" the human stock. The eugenics movement was especially popular between the two World Wars, giving rise to sterilization laws in 24 U.S. states, and restricting "unwanted" immigration.

In its most extreme expression, eugenics was used by Nazi Germany, primarily against incarcerated Jews before and during the Second World War. Although it soon fell out of favor, it is being revived thanks to new technologies whereby human engineering promises a cocktail of genetic possibilities.

Although eugenics is an "artificial" application of Darwin's natural selection, there are contemporary Darwinists who argue for its use to "improve" society. Among them are Nobel laureate James Watson, who argues for selective genetic reengineering of human beings, and Richard Dawkins, who commented that Hitler gave eugenics a bad name.

The Least You Need to Know

- The Enlightenment period and the age of reason are the same.
- The Enlightenment was a philosophical movement.

◆ The scientific method began to take on a new look, with little room for God.

◆ William Paley believed the universe resembled the design of a watch.

◆ Darwin's "survival of the fittest" bore social implications wherein all men were not created equal.

God, Science, and American History

In This Chapter

- ◆ The great awakenings
- ◆ A watershed trial
- ◆ Modern Intelligent Design

While the Enlightenment was in full swing in Europe, a much younger America was swept up in a religious culture that set the stage for significant friction between science and religion.

Of course, the eighteenth-century evolution revolution wasn't happening in a vacuum. To this day, the United States is known to have one of the most religious cultures in the Western world. The past is no exception, beginning with the Pilgrims' quest for religious freedom, which would propel the culture through frequent secular waters into the twentieth century.

America's "Christian" Beginnings

America's religious history may be seen as being composed of three "great awakenings." Each one lasted about a hundred years, or three generations. They tended to overlap, and each was begun by a phase of religious revival, inspired by social and technological changes that outpaced the prevailing moral vision.

In the face of the ethical and moral crises such changes inevitably imposed, religious adherents had to reconstruct a new theological vision. This was naturally followed by a political reform that would realize the new vision—before the political coalition sustaining these reforms would go into decline. In turn, a new generational crisis would give rise to a new "awakening," and so the cycle has continued through our history of no less than three hundred years.

The First Great Awakening

We've all read in history books how it was religious freedom that was the motivating force behind the Pilgrims' journey and the birth of our nation. It is difficult now, in the midst of the religious freedom we presently enjoy, to imagine how threatened, and therefore how precious, this liberty was to them. Yet the eighteenth-century evangelicals, commonly known as "new lights," were the first to "awake" and the first to critically shape religion in America.

When Jonathan Edwards appeared on the scene in the 1730s and 1740s, religious faith had been in decline for several decades. The Enlightenment, with its convincing message of the power of human reason, had eroded the need for dependence upon, or even recognition of, God. The Salem witch trials further discredited this early Christianity, leaving a void that would soon be filled by Edwards.

Evolutionary Revelations

It is interesting to note that Jonathan Edwards's Northampton church, said by Yale University to have been "the largest and most influential church outside of Boston," now struggles to survive, lurking on the edge of a vibrant cultural and economic boomtown.

At his church in Northampton, Massachusetts, Edwards used his rhetorical power to preach hellfire, sin, and salvation. "Sinners in the Hands of an Angry God" was a sermon typifying his zeal, inspiring New England, just as other colonies were inspired by other like preachers.

In response to the newborn opportunities afforded by colonial life, this religious rebirth was driven by a culture of economic opportunity. Emphasizing an egalitarian Christian spirit that stressed common perseverance, the groundwork was laid for each individual to achieve his personal salvation. Its conflict with the Enlightenment assumption of impersonal human reason would set the stage for a coming battle with Darwinian scientific thought.

The Second Great Awakening

John Calvin, the great Reformation theologian from Picardie, France, believed in "predestination": the idea that before a person's birth, God had already predetermined whether or not he would be saved.

In the midst of a prevailing cultural spirit of personal opportunity, Calvin's doctrine of doom made less and less sense, requiring a new theological vision. It was the newborn emphasis on the possibility that anyone could be "chosen" that inspired individuals to commonly prepare for "the second coming of Christ."

Fact or Faith _____

John Calvin's unyielding belief in predestination frightened many believers who feared that, despite their own Christian faith, their soul might already be damned.

The second great awakening marked a growing split in a formerly more unified culture. The secularism of the Enlightenment continued to build in the land, inspired by such great Enlightenment thinkers as Benjamin Franklin and Thomas Jefferson. This deistic thinking, which was skeptical of "personal religion," would challenge the special place of human beings and the promise of personal salvation.

Enter Charles Darwin. With its chance-driven explanation of human life, without the need for a God, evolutionary theory was perfectly positioned to offend religious people. After all, Darwin's contention that life operated on its own not only offended believers by its implied atheism; it undermined their sense of themselves. In Darwin's scheme, man was no longer created in the image of God. He was a chance accident, even an afterthought in the shadow of natural selection.

Evolution dealt a fundamental blow to American religious faith—not only for God's absence by its "self-sufficiency," but for its outright defiance of the Bible. The creation story of Genesis was denied by Darwin's contention that human beings descended from a common ancestor, rather than from Adam and Eve. And by Darwin, humans lost their "special" place that gave purpose to their daily lives—and whose significance was no longer as God's unique gift, but as a link in a godless chain of being.

The Third Great Awakening

The Origin of Species, published in England in 1859, was lost on American culture in the midst of the Civil War and its fallout. Gradually, however, Christians would grasp its profound implications for their lives, which challenged the infallibility of the Bible—that for many was the bedrock of their faith. The battles that ensued over interpreting the Bible were already going on, but Darwin was becoming a scientific priest who had concrete "proof" on his side.

Along with the philosophy of the Enlightenment came the industrial revolution, which increasingly demanded the wisdom of science for its competitive success. The turning of the twentieth century demanded a vision that could somehow accommodate its own unfolding knowledge and experience. A new generation was rising up—of Christians as well as non-Christians—who were beginning to ask unprecedented questions that were soon to be answered by science.

It is important to note that this was also the age of Social Darwinism, with its unqualified materialism and doctrinal opposition to traditional Christianity. In its shadow, "modernist Christianity" attempted to reconcile religious experience with science; secular humanism de-emphasized God in favor of human triumph; agnosticism and atheism often took the form of socialism and communism; and in response, fundamentalist Christianity dug in its heels and held tight to its Bible.

The Scopes "Monkey" Trial

The Scopes "monkey" trial is one of the most celebrated trials in our history. It is significant not only because of its religious and constitutional implications, but also because it became a Hollywood phenomenon that continues to speak volumes about the time and culture in which it took place.

In 1925, the inconsequential town of Dayton, Tennessee, became an overnight stage on which the battle would be waged between religion and thinking man's science.

More than 100 reporters from the United States and Europe were dispatched to the Tennessee hills. Each day of the trial, 22 telegraphers tapped out 165,000 words, thus dominating the national media with this astonishingly landmark case. Movie film was flown out every day from an airstrip created for that purpose, and *Inherit the Wind*, a Broadway play, would be produced about the extravaganza.

The American Civil Liberties Union had offered to defend anyone who had been indicted for teaching Darwin's theory of evolution in a public school. A local businessman, seeing opportunity in bringing this contentious case to town, lobbied other businessmen for their support in approaching local resident John Scopes. Scopes was a high school football coach who had substituted in a science class—and who, by virtue of having used an evolution textbook, had defied the Butler Act.

The Butler Act

In the 1920s evolution was a controversial scientific theory—even among scientists, and surely among Christians who saw it as a road to atheism. William Jennings Bryan, former secretary of state and presidential candidate, still enjoyed significant political influence in Tennessee, as well as in the nation. His book *In His Image*, asserted that evolution was irrational and immoral, stirring up the public and the Tennessee legislature, which ultimately passed the *Butler Act*.

def•i•ni•tion

> According to the **Butler Act**, "… it shall be unlawful for any teacher in any of the Universities, Normals and other public schools of the State which are supported in whole or in part by the public school funds of the State, to teach any theory that denies the story of the Divine Creation of man as taught in the Bible, and to teach instead that man has descended from a lower order of animals."

It was this act that the ACLU wanted to challenge, and did so behind noted attorney, Clarence Darrow. The state of Tennessee required the use of the textbook *Civic Biology*, which, because it taught evolution, effectively demanded that public teachers break the law. Although John Scopes couldn't recall having taught the evolution section, he went along with the drama.

Paradoxically, Bryan's opposition to the textbook in part had to do with its teaching of eugenics—the belief that the human gene pool can be, and should be, improved through selective breeding. A portion of the book was written by Charles Davenport, who was Director of the Eugenics Records Office, a privately funded research organization. The text was perceived to argue for the inherent superiority of the white race, and advocated a eugenics-oriented policy to eliminate the "genetically inferior" members of society.

Evolutionary Revelations

Although the episode, made famous in the 1960 film *Inherit the Wind*, is often seen as a triumph for progress and science, the textbook at the heart of the court case, *Civic Biology*, has a white supremacist and eugenics bias. It states: "We do have the remedy of separating the sexes in asylums or other places and in various ways of preventing intermarriage and the possibilities of perpetuating such a low and degenerate race. Remedies of this sort have been tried successfully in Europe and are now meeting with success in this country."

—Donald MacLeod, *The Guardian* (U.K.), September 26, 2005

A New Religious War

The trial had the tone of a religious war from both sides of the aisle. Whatever the objective merits of the case, the science was secondary. Theologian Paul Tillich defines religious faith as a matter of "ultimate concern": the beliefs about which one is ultimately concerned are, de facto, religious beliefs. The tone and strategies employed in the trial expressed an urgency that seemed to manifest a religious fervor well beyond the legal case itself.

Thus Bryan would indict Darrow's past legal defense of two child murderers as a sign of his personal immorality, and so that of evolution. Reciting Darrow's own words, Bryan attacked: "This terrible crime was inherent in his organism, and it came from some ancestor ... Is any blame attached because somebody took Nietzche's [evolutionary] philosophy seriously and fashioned his life upon it? ... It is hardly fair to hang a 19-year-old boy for the philosophy that was taught him at the university."

In response to Bryan's religious rebuttals, Darrow would say: "You insult every man of science and learning in the world because he does not believe in your fool religion."

Evolutionary Revelations

The famous journalist, satirist, and critic H. L. Mencken brought prominence to the case with his brilliant use of exaggeration and vivid embellishment. Aligned with Darwinian evolution, Mencken called Dayton residents "morons" and "yokels," and Bryan, a "buffoon" spouting "theologic bilge." When referring to the legal defense, Mencken praised the lawyers as "eloquent" and "magnificent"—and by some, was given credit for having turned public favor against creationism.

There is a haunting quality about the present controversy over Intelligent Design—giving one to wonder if, as Darwinists complain, we've progressed at all since then? The answer may be no, in that it remains a conflict between two "ultimate concerns." Yet the answer may be yes, in that Intelligent Design poses new questions that have to be answered.

In that, let's look at the recent history of ID. We say "recent," because the idea has existed—if not in name—as a concept for centuries.

The "New" Intelligent Design

Jonathan Witt, a senior fellow at ID's think tank, The Discovery Institute, reminds us that ID critics tend to lump Intelligent Design with creationism. They assert that when the teaching of creationism in the public schools was struck down by the Supreme Court in 1987 (*Edwards* v. *Aguillard*), this "stealth creationist" movement was begun in an attempt to subvert the law. However, Witt contends, although Intelligent Design has its roots in ancient Greece, even the recent ID movement predates this Supreme Court decision.

Renowned chemist and philosopher of science Michael Polanyi once noted that just as there are machines that cannot be reduced to physical laws, so certain organic structures seem to exist as "irreducible" wholes. With the advent of molecular research in such complex information systems as DNA, physical laws seemed inadequate to describe their complex function. DNA expressed a complex design beyond a blind material process.

ID's American Launching Pad

Witt attributes Michael Behe's idea of "irreducible complexity" (a term you will get to know well) to Polanyi's early work. But it was the publication of *The Mystery of Life's Origin* in the 1980s that first fueled the movement that would become known as "Intelligent Design." Scientists Charles Thaxton, Walter Bradley, and Roger Olsen authored this catalytic work, suggesting the possibility of metaphysical "design" as a "scientific explanation."

It is interesting to note that until the popularization of Intelligent Design—that is, until "design" was seen as a stealth, religious, creationist movement—Thaxton's book was extremely popular in academic circles, and widely accepted for its scholarship. Incorporated into The Philosophical Library of New York, which has published many

Evolutionary Revelations

Charles Thaxton did not use the term Intelligent Design until he edited *Of Pandas and People*—a textbook exploring the question of biological origins.

Nobel laureates, *The Mystery of Life's Origin* "became the best-selling advanced college-level work on chemical evolution," according to Jonathan Witt in his article "The Origin of Intelligent Design: a brief history of the scientific theory of intelligent design." Yet in today's anti-ID academic climate, it is hard to imagine its positive review in the *Yale Journal of Biology and Medicine*.

Witt explains: "As it [the book, *Of Pandas and People*] neared completion, Thaxton continued to cast around for a term that was less ponderous [than creative intelligence, intelligent cause, artificer, and intelligent artificer] and, at the same time, more general, a term to describe a science open to evidence for intelligent causation and free of religious assumptions. He found it in a phrase he picked up from a NASA scientist. 'That's just what I need,' Thaxton recalls thinking. 'It's a good engineering term … After I first saw it, it seemed to jibe. When I would go to meetings, I noticed it was a phrase that would come up from time to time. And I went back through my old copies of Science magazine and found the term used occasionally.'"

In fact, it is hard to imagine Intelligent Design ever being an innocent term. As politically charged as it is today, one envisions a sinister plot. Indeed, many critics see the term and the movement as no less than a conspiracy. Some see it as an effort to sneak God into the schools or to evangelize the nation; more scientific critics are inclined to simply view it as "bad science."

A "Scientific" Journey?

To hear Intelligent Design proponents tell it, ID's history is not a story of "masked creationism" but a story of a scientific journey. In fact, they cite numerous supporters who have little religious conviction, or who believe in the presence of material design but don't want to be "ID-entified." Yet mainstream scientists look to such agencies as the Discovery Institute as proof of ID's political agenda.

Among such non-ID scientists is information theorist Hubert Yockey, who contends that the intricate DNA sequences are mathematically identical to written language, and so imply an "intelligence." British philosopher Anthony Flew, who made a name by his atheism—entering into numerous hostile debates on the subject of Intelligent

Design—famously turned from his skepticism (though still rejecting a Christian God) to assume a deistic position that entertains transcendent design. And renowned astronomer Sir Frederick Hoyle, who rejected the idea of God, saw evident design in the universe across the scientific disciplines.

"A commonsense interpretation of the facts suggests that a superintellect has monkeyed with physics, as well as chemistry and biology, and that there are no blind forces worth speaking about in nature," writes Paul Davies in *The Accidental Universe*.

Evolutionary Revelations

In my own conversation with Michael Behe, I asked him how much contact he had with other ID "players": Michael Denton, William Dembski, Phillip Johnson, and indeed, the Discovery Institute. I had imagined they were all in constant contact (perhaps thanks to the media, but also given that most of them were "Senior Fellows" of the Discovery Institute). He explained that he'd seldom met with them in person but that with some he was in e-mail contact. In an interview about his political role in the controversy, he once said that he preferred drinking beer and mowing the lawn.

—Author Christopher Carlisle, March 11, 2006

A Unique Approach

As with any heated conflict where a reigning paradigm is being challenged from the outside, there are bound to be misleading and inaccurate portrayals of the motives of the challenger. But as we shall see, the controversy is far more interesting than the media has conveyed amidst the cultural battles, the court fights, and the academic wars.

The program of Intelligent Design is to make, at most, "minimalist claims," in which no attempt is made to exceed design detection, to identify any designer.

As ID proponent Jonathan Witt says, "Consider intelligent design's most famous design inference, the bacterial flagellum. Michael Behe shows that this microscopic rotary engine, like an automobile engine, needs all of its machinery in place to function at all. The best explanation for this irreducibly complex machine is intelligent design, but there's no inscription on the bushing of this little motor that identifies its maker. To discover the identity of its designer(s), one has to look beyond science."

The Least You Need to Know

♦ There were three "great" religious "awakenings" in American history.

♦ The Butler Act legally forbade any teacher in any Tennessee public school to teach any theory denying the story of the divine creation as taught in the Bible.

♦ The Scopes trial pitted "creationism" against science in a 1920s Tennessee courtroom. Creationism lost.

♦ Charles Thaxton coined the term Intelligent Design.

♦ According to Intelligent Design proponents, ID is not "masked creationism" but a scientific enterprise.

Part 3

"Traditional Science" Applications

The physical world is currently perceived by science's "big three" disciplines—physics, chemistry, and biology—as processes of evolution. And each, as you will see discussed in this section, plays a major role in the ongoing debate over the theory of Intelligent Design.

Here we examine each discipline, and its impact on the ID debate.

Chapter 8

Mainstream Physics

In This Chapter

- ◆ Universal beginnings
- ◆ Smaller than an atom
- ◆ Carter's principle
- ◆ Perfection explained

Creation is, of course, a central focus of Intelligent Design, and no one on either side of the debate is pretending otherwise. In fact, as we already have seen, ID grew in part out of the creationist movement, which was concerned with the importance of life's beginnings as a window through which to see God.

Whether or not we believe in God, since the advent of self-consciousness human curiosity has moved us to ponder our origins: "from whence we came."

Moreover as we shall see in the next few chapters, there is dramatic disagreement between mainstream science and Intelligent Design about the nature of the physical world.

Of the various scientific disciplines, physics seems to resonate with theistic understandings of the universe more than chemistry or biology does. Still, the vast majority of mainstream physicists assert they do not believe in a God, and they have generally developed atheistic understandings of our physical world. Unlike Intelligent Design proponents, mainstream physicists generally limit themselves to physical interpretations of the universe (indeed, even when they do believe in God).

Fact or Faith _____

Generally speaking, the disagreements between mainstream science and Intelligent Design have less to do with experimental results than with their interpretation of what the experimental results ultimately mean. As with any information, scientific information cannot speak for itself; it is a tool to be used in the process of human understanding.

The Big Bang

Until the big bang theory, physicists presumed a "steady-state" universe: that is, a universe which always "was" and always "would be" according to certain unchanging constants. It implied an unbounded or infinite space, which was simply "a given" within which everything existed. The planets, the stars, the galaxies, and our island home, were contained in a vast, never-ending context. It was as though space (and time) were to be treated as a passive playing field on which the universal game was played: given brief attention as a necessary backdrop for the movements of its physical contents.

It was with the general theory of relativity in 1917 that Albert Einstein took a different route. Einstein said that the universe was not a static space, but a dynamic reality. He understood that if the universe were static, the galaxies should collapse into each other. Thus he established the notion of a "cosmological constant" to explain their suspension within space.

But in 1920, astronomer Edwin Hubble measured the light from distant galaxies to indicate that the universe was expanding. Einstein could abandon his "cosmological constant"—which he came to see as a mistake—and accept the revolutionary idea that the universe itself was growing bigger. In 1927, another astronomer, Georges Lemaitre, developed the first model of what has become known as "the big bang" theory of creation, which could explain such an expansion.

The Beginning Explained

The big bang marked the creation of the universe rather than a creation within the universe.

So steeped are we in the "steady-state" model, where time and space are seen as "a given," it is difficult to grasp the radical idea that they both came from a singular source. What this means is there was actually a physical creation which happened at "time zero"—a point where the universe came into being, before which, as with time there was no space. Einstein's general theory of relativity describes this space-time universe.

Using Einstein's theory, physicists since have calculated the universe's age, and by observing the heavens, have come to believe that the universe is indeed expanding.

Blinded By Science _____

Physicists have calculated the universe's age by observing "red shift" light, which indicates physical movement away from the observer (not unlike the Doppler effect, by which a horn pitch falls as its train speeds by the railway station where you are standing). They retraced the trajectories of heavenly bodies receding from their telescopes, and thereby deduced the universe is roughly 13 billion years old.

Explosion of Time and Space

Scientists could further deduce that there was a unique point from which (or into which) the universe came. Rather than there being a given empty space, in which stars and galaxies always existed, space itself was once infinitely compressed, exploding forth as time and space at the moment of the big bang. This initial state is known as a "singularity"—the first instance of time and space from which all stars would form: comprising galaxies and solar systems like our own.

By calculating temperatures and background radiation with the use of powerful telescopes, cosmologists calculate the universe had to have begun as a super dense fireball. Since then, the universe has been gradually cooling. And this cooling from a super-hot state has finally allowed life to emerge as we know it. To some this singular event may appear to be a divine "miracle," whereas for others, specifically, most physicists, it is scientifically understandable.

A Beginning with No Beginner

Although there are many variations of the big bang model (perhaps as many variations of the model as there are physicists), the theory currently enjoys overwhelming acceptance in the scientific community. Like most "successful" theories, it most successfully explains a spectrum of phenomena, and its critics will have to reckon with its convincing explanations in order to supercede the theory.

Blinded By Science _____

No one knows exactly how the first space, time, and matter arose. And scientists are grappling with even deeper questions. If there was nothing to begin with, then where did the laws of nature come from? How did the universe "know" how to proceed? And why do the laws of nature produce a universe that is so hospitable to life? As difficult as these questions are to answer, scientists are attempting to address them with bold new ideas and experiments to test these ideas.

—NASA and the Harvard Smithsonian Center for Astrophysics, 2004

The existing evidence for what caused the big bang suggests that there was a universal singularity which started it all off. Of course this begs the question: doesn't there have to be a beginner?

One common response has to do with the nature of the singularity itself. World renowned physicist Stephen Hawking points out that the singularity (at time equals zero) is where physical laws do not apply—and can't until the point of creation. To impose an order made of present laws defies the lawless order that reigned at the moment of "the bang;" so any attempt at a rational explanation must recognize this limitation.

Thus at the start, mainstream physicists argue, chaos necessarily prevailed: that by the nature of the big bang's lawlessness, determined outcomes were impossible. It wasn't that the laws had yet to take hold of the chaos of the fiery explosion, but rather that these laws were in the process of formation and had yet to be created. Therefore, one could say, anything could have come out of the fiery explosion; it just so happened that this particular universe came out—by which we're here to tell about it.

Quantum What?

One might argue that a remnant of the chaotic big bang is the uncertainty of something we call "quantum mechanics."

Sounds complex: so what is it? Quantum mechanics is a physical theory that deals with phenomena at the subatomic (smaller than an atom) level in more successful ways than does Newtonian mechanics. It was developed to explain how, as an example, an atom's negatively charged electron stays in orbit around a positively charged nucleus, instead of collapsing into it.

Elusive Electrons

Classical electromagnetic theory would predict such a collapse, whereas quantum theory explains that electrons can't be conceived as bounded particles. They must rather be seen in the context of a diffuse or scattered "cloud" bearing a negative charge, which can only be defined in a field of probabilities dispersed around their orbits. It is important to note that according to the theory, this lack of "clarity" isn't a matter of inaccurate measurement; particle locations are inherently uncertain, and can never be statically determined.

The Heisenberg Uncertainty Principle

The Heisenberg uncertainty principle is a cornerstone of quantum physics theory.

Werner Heisenberg found that it was impossible to accurately measure both the location and momentum of an individual particle: when one has measured a particle's location, its momentum becomes unknowable; when its momentum is determined, its location becomes blurred. Thus at the subatomic level, nature can be probabilistically described but never defined with absolute certainty.

This departure from determinism has upset scientists and nonscientists alike. In fact, Einstein resisted the idea, despite the fact that he helped to father quantum mechanics. He said, "I cannot believe that God would choose to play dice with the universe." He expressed our common hope that armed with enough knowledge and know-how anything can be known.

The point of quantum theory is not to say that nothing can be known. It is to say, however, that elementary nature must be known probabilistically. Here no amount of knowledge can surmount the inevitable uncertainty of the universe.

Quantum uncertainty has significant theological implications. For those who believe in God as an immutable, deterministic ruler, to speak of uncertainty is to question God's power, and even, perhaps, God's existence. Most mainstream

> **Evolutionary Revelations**
>
> German physicist and Nobel prizewinner Werner Karl Heisenberg was a key figure in the development of quantum mechanics. He was also a member of the scientific team that was working to develop an atomic weapon for the Nazis during World War II, though just how deeply he might have been involved is the subject of conjecture.

> **Evolutionary Revelations**
>
> Niels Bohr, a colleague of Albert Einstein and a cofounder of quantum theory, is said to have told Einstein, "Don't tell God what to do."

physicists respond that uncertainty is simply and inherently the way of the world: both by how it was created, and how it is sustained in its seemingly accidental nature.

The General Anthropic Principle

In a paper given in 1973 by Cambridge astrophysicist and cosmologist Brandon Carter, the "anthropic principle" was introduced to describe the apparently uncanny way in which the universe seemed created for human existence. In the face of the many physical constants that are present in the universe, the one thing these constants seem to have in common is their hospitality to "life." Carter pointed out the myriad physical laws that conspired to "create" the universe, and that continue to sustain our physical world as we go about each day.

The Force of Gravity

Gravity is about 1,039 times weaker than the electromagnetic force. If gravity had been a little weaker relative to electromagnetism (1,033 to be exact), stars would be a billion times less massive and would burn a million times faster. As a result, there would not have been enough time for stars to create heavy elements such as carbon, which is essential for the creation of life as we know it.

Strong and Weak Forces

If the nuclear strong force, the force that holds together the particles of an atom's nucleus, were just slightly weaker than it is, nuclei with several protons would not hold together, and stars would not be able to form. Moreover, hydrogen would be the only element in the universe. If the nuclear strong force were 1 percent stronger, hydrogen would be extremely rare in the universe. In either case, life as we know it would be impossible.

The nuclear weak force affects the behavior of subatomic particles (things such as neutrinos, electrons, and muons). If this force were slightly greater, little helium would have been produced by the big bang. As a result, heavy chemical elements necessary for the creation of life could not have been cooked by the stellar furnaces. And if the force had been slightly less, all the hydrogen would have been turned into helium. Among other things necessary for life, water would have never been created.

The relative masses of these three particles seem perfectly engineered. A neutron's mass is a little bit more than the combined mass of the proton and electron, allowing a neutron to decay into a proton, an electron, and a neutrino. If the neutron's mass had equaled the combined masses of the proton and electron, then hydrogen, which is necessary to fuel the stars, would have never been produced. Again, life as we know it would have been impossible.

The Perfection of Water

We all know that water is absolutely necessary to human life. What we may not know is that water molecules possess certain properties that are unusual among the elements. Specifically, water is lighter in liquid than it is in solid form. Because ice floats, the earth's water doesn't freeze from the bottom. If it did, the earth would have been covered in ice, ending life as we know it.

It should be noted that there are other manifestations of the anthropic principle. The creation of carbon, which is the fundamental element of organic life, is a product of a seemingly astonishing ratio between the strong force and electromagnetism. Entropy—the measure of heat that is absorbed by a system—makes the universe seem finely tuned to create stars and solar systems necessary to create and sustain human life.

So the scientific question is, are these just coincidences? Is there a coherent explanation? How might one look at and assess these conditions from a scientific point of view?

The Weak Anthropic Principle

The "weak" version of the anthropic principle (the principle that the universe seems uniquely and perhaps intentionally hospitable to life) is more popular with mainstream physicists than the "strong" version.

The weak anthropic principle essentially adopts a "So what?" attitude toward the universe. Unlike the strong version, which is favored by proponents of Intelligent Design, the weak anthropic principle essentially says: "What is, is simply what is." That's that. Don't worry about it.

In his book *A Brief History of Time*, physicist Stephen Hawking expresses the common scientific view that the universe does little more than prove itself.

Blinded By Science

One example of the use of the weak anthropic principle is to 'explain' why the big bang occurred ten-thousand million years ago—it takes about that long for intelligent beings to be involved … an early generation of stars had to form. These stars converted some of the original hydrogen and helium into elements like carbon and oxygen, out of which we were made. The stars then exploded as supernova, and their debris went to form other stars and planets, among them those of our solar system, which is about five thousand million years old. The first one or two thousand million years of the earth's existence were too hot for the development of anything compli-cated. The remaining three-thousand-million years or so have been taken up by the slow process of biological evolution, which has led from the simplest organisms to beings capable of measuring time back to the big bang.

—Stephen Hawking, *A Brief History of Time*

In other words, to say that the universe is finely tuned for human life is to create a "tautology": a statement that is already true by its own definition. In this view, the outcome is already rigged before science could get started; it is like marveling that every triangle one sees has exactly three sides. Here, the universe is finely tuned for human life because if it weren't, we wouldn't be here to see it.

Explaining Perfection

The term *finely tuned universe* comes from the idea that there is a seemingly miracu-lous constellation of physical constants that conspire to allow human life to exist.

"Fine-tuning" is an expression of the anthropic principle that focuses upon the physi-cal constants themselves. Though some scientists believe in naturalistic sources of fine-tuning, others write off "fine-tuning" as a product of random chance or the inevi-table product of a "multiverse."

A Roll of the Dice?

The argument for random chance simply states that through the millions of pos-sibilities, the creation of our universe is just what we happened to get. Stephen Jay Gould belonged to this school of thought, holding to the idea that chance rolled the dice, and it happened to give us a "double six." However lucky the result happened to be, the roll had to produce something, and so we shouldn't be amazed that we ended up with the universe we have.

Mainstream physicists point out that winning a million-to-one lottery is unlikely enough that any individual would be foolish to count on it. Yet someone has to win, the argument goes, and in this case "the winner" was us.

A "Multiverse"

Another way to understand "the miracle" of our universe is to entertain the possibility that we are simply one universe among many. If there is a whole host of possible universes, then there is no reason to assume that ours wouldn't exist with its own particular laws. In this mainstream scientific perspective, to say that we are "special" is blindly "anthropomorphic": that we are blindly centered on ourselves, and are ignoring the myriad possibilities of physical creation.

Philosopher Quentin Smith elaborates on this perspective, while asserting the universe's inherent richness. He says that our world "exists non-necessarily, improbably, and causelessly. It exists for absolutely no reason at all. It is inexplicably and stunningly actual … The impact of this captivated realization upon me is overwhelming. I am completely stunned. I take a few dazed steps in the dark meadow, and fall among the flowers. I lie stupefied, whirling without comprehension in this world through numberless worlds other than this one."

To the extent that mainstream physicists recognize the apparent "fine-tuning" of the universe, this is generally explained within the material nature of the world itself. Mainstream physics, as it is presently conducted, seeks physical explanations, rejecting metaphysical answers as finally inadequate.

Cosmologist, Sir Martin Rees, represents such a view: "People used to wonder: why is the earth in this rather special orbit around this rather special star, which allows water to exist or allows life to evolve? It looks somehow fine-tuned. We now perceive nothing remarkable in this, because we know that there are millions of stars with retinues of planets around them: among that huge number there are bound to be some that have the conditions right for life. We just happen to live on one of that small subset. So there's no mystery about the fine-tuned nature of the earth's orbit; it's just that life evolved on one of millions of planets where things were right."

The Least You Need to Know

- Until the big bang theory, physicists presumed the universe to be in a "steady state" of time and space.

- The big bang marked the creation of the universe rather than a constantly existing universe.

- Quantum mechanics is a theory that deals with phenomena at the smaller-than-an-atom level, and attempts to explain particle behavior.

- Quantum mechanics says that nature can only be known probabilistically.

- The general anthropic principle attempts to explain how the universe "appears" created.

Mainstream Chemistry

In This Chapter

◆ The chemistry of ID

◆ Aristotle's view of life's beginnings

◆ Life just happens?

◆ DNA tells quite a story

Chemistry! The mere mention of the word evokes feelings of test-taking anxiety … for good reason. The study and understanding of chemistry is not easy. Many of us struggled with it in high school, and hoped never to deal with the subject again. But if we are going to wrap our minds around the Intelligent Design debate, we must first have a basic grasp of chemistry.

So let's look at in a simple way.

Chemistry is the study of the structure and properties of physical matter. It focuses on how physical elements interact with one another. And it has, in recent years, become extremely more complex because technology has enabled chemical experiments that were never possible before. But chemistry is also more fascinating today, because with the modern world's capability of analyzing things at the atomic and subatomic (things smaller than an atom) levels, many mainstream chemists feel confident that they are on the threshold of answers to mysteries which have baffled us from the time of Aristotle.

Chemistry has over the centuries been occupied with the question of how physical matter might move from inorganic matter (nonliving stuff) to organic life (living things). As chemical dynamics at a molecular level have been revealed by modern research, many mainstream chemists believe they are on the verge of unlocking the secrets of how physical matter can interact to both support and develop biological life. Indeed, the discovery of the structure of DNA by Francis Crick and James Watson in the 1950s is seen not only as one of the great scientific revelations of the twentieth century, but as a way to understand the very language of human life itself.

ID, Chemically Speaking

As we have seen, much has been made of Intelligent Design's conflict with mainstream science in the field of physics—and particularly, in the realm of cosmology. Moreover, because of its frequent political battles with Darwinian evolution, Intelligent Design has often focused on biology. But the mainstream science-ID conflict extends equally to chemistry, where the question of design is now being argued at the molecular level.

As we are about to discover, Intelligent Design views the chemistry of life as a function of "fine-tuning," which enables life to thrive and for us to be around to observe it. Yet as one might predict, mainstream science rejects fine-tuning arguments. Going back to the "weak anthropic principle," mainstream science is apt to say the stuff of life simply "is as it is"—and all we need to do is understand it.

 Blinded By Science _____

Physicist Victor J. Stenger describes the weak anthropic principle as it is applied to the chemistry of life: "The observed values of all physical and cosmological quantities are not equally probable but take on values restricted by the requirement that there exist sites where carbon-based life can evolve and by the requirement that the Universe be old enough for it to have already done so."

He continues, "The WAP has not impressed too many people. All it seems to say is that if the universe was not the way it is, we would not be here talking about it. If the fine structure constant were not 1/137, people would look different. If I did not live at 508 Pepeekeo Place, I would live someplace else."

Mainstream scientists may concede that the universe "appears" to be fine-tuned; much of their disagreement with Intelligent Design isn't about the material stuff. It has rather to do with its interpretation—whether it truly is "fine-tuned"—and therefore, whether or not there is a "Tuner" who intervened (or intervenes) in the world. If indeed life simply is as it is, then mainstream chemistry must simply go about exploring the chemical elements just as they are given to us.

Life-Proving "Carbon"

When we think of carbon, most of us may think of pencil lead, or perhaps of diamonds.

Carbon assumes several "allotropic" forms—or molecular constellations—that allow it to be soft, as in the case of graphite, or hard, as in the case of dazzling gems. Its ability to bond with itself and a host of other chemical elements enables it to form more than 10 million different compounds known in the universe. Its uses are prolific: from radiocarbon dating to the manufacture of steel, from use in nuclear reactors to making charcoal for cooking, and a host of vital medicines.

Carbon is present throughout organic life. When it binds with oxygen, carbon dioxide is produced to sustain plant life on earth, while bonding with hydrogen produces hydrocarbons which manufacture fossil fuels. When oxygen and hydrogen are added to the mix, we have a recipe for human sustenance; creating fatty acids, and even adding flavor to the fruit we eat every day.

It Wasn't Always Here

We know that carbon wasn't created in the big bang: it was more gradually cooked in the interiors of stars than the big bang would allow. Because it is found in great abundance in the bodies of our solar system, it is readily available to be flexibly used in a million life-sustaining ways. In fact, carbon is so critical to life—it is hard to imagine life without it—that one can easily understand how carbon is construed to be "perfect" for human existence.

But this is the point for mainstream scientists: it is perfect for *this* human existence. If it is true that carbon is uniquely suited to the many requirements for life, perhaps only for this life, and not to other life which would require a different chemical basis. In the spirit that physicists argue for the likelihood of other universes, mainstream chemistry might see these universes as made of very different stuff.

Perfect and Essential Water

Water is so basic to human existence that it is apt to be overlooked. Tasteless, odorless, and even transparent, it almost seems as if it were made to go unnoticed.

About 72 percent of the fat-free mass of the human body is water, and without it, human life could not evolve, and each of us would die in several days. Water causes floods, destroying human populations, and water's absence does the same; water makes for industry, vital transportation, even leisure entertainment.

At the molecular chemical level, water is deceptively simple. Constituted by two hydrogen atoms that are attached to two oxygen atoms (actually H_2O_2, not H_2O), its dynamic properties make it as essential to human existence as carbon. Morphing from liquid, to solid, to gas, water bears the odd characteristic of being lighter when it's solid—due to its crystalline molecular configuration. If ice were heavier than its liquid counterpart, water would freeze from the bottom, making the oceans single blocks of ice and prohibiting life on Earth.

Water enables organic compounds to react for cellular replication, allowing life to reproduce. Absorbing infrared radiation, water plays a major role in the atmospheric greenhouse effect by which a habitable temperature is maintained, instead of one far below freezing. And due to its qualities by which it is continuously exchanged within the hydrosphere, water sustains life underground, on the surface of the earth, and in the air.

Simply put, water is essential.

The Air We Breathe

The earth's atmosphere consists of a blanket of gases that covers the entire planet. Composed of 78 percent nitrogen and 21 percent oxygen (and many other gases), "air," as we have come to know it, serves many critical functions.

Like water, it protects a fragile human life from the assaults of ultraviolet radiation, and stabilizes the earth's temperature from extremes imposed by day and night.

Brief History of Atmospheres

Although there will always be disputes in theoretical reconstruction of the development of the earth, one widely held scenario is that there have been three historic atmospheres over the course of our planet's existence. Each atmospheric stage represents a different chemical composition. The first persisted until 3.5 billion years ago, while the earth was still cooling. Mostly constituted by helium and hydrogen, it was gradually dissipated by the heat that emanated from the molten crust and by radiation from the sun.

Blinded By Science

The air surrounding the earth consists of a series of atmospheric layers: the troposphere in which we live, then the stratosphere, the mesosphere, and the thermosphere. Each layer has its own atmospheric pressure, as pressure depends upon air's weight. So most of the atmosphere huddles near the earth—more than half below the peak of Mount Everest—and is more or less composed of a uniform chemical composition, called the homosphere.

As the earth's surface cooled to form a crust, it was still bubbling with volcanoes: releasing steam, carbon dioxide, and ammonia, and creating a "second atmosphere." Carbon dioxide and water vapor were its principal chemical composition; containing almost no oxygen, it couldn't support the most primitive organic life. For several billion years, the water vapor condensed into rain, making bodies of water and dissolving the carbon dioxide (half of which would be finally absorbed by the oceans).

It is believed that about 3.3 billion years ago, the first oxygen-producing organisms emerged. They would convert the current atmosphere to one with oxygen and nitrogen through *photosynthesis*—a process that would later be assumed by plants evolving in this transforming atmosphere. As these plants continued to oxygenate, carbon dioxide levels dropped, and oxygen amassed in the atmosphere to change life once again. Life forms dependent upon carbon dioxide were followed by oxygen-friendly forms, giving rise to the evolution of our own "third atmosphere" composed of oxygen and nitrogen.

def•i•ni•tion

According to *Merriam-Webster*, 2006, **photosynthesis** is "the synthesis of chemical compounds with the aid of radiant energy and especially light; especially: formation of carbohydrates from carbon dioxide and a source of hydrogen (as water) in the chlorophyll-containing tissues of plants exposed to light."

As remarkable as these elements—earth, air, water, and so many others—may seem, the job of mainstream chemistry is to understand them as the basic stuff of nature. Science must look at the chemical material that has been given in the universe and reconstruct, by "best explanation," what has gone before, and what is yet to come. If there is design in the universe, such chemistry isn't interested; it is more than enough to explore nature itself and give natural explanations.

"Life's Origins," to Aristotle

Although many ancient philosophers have put their "oars" in the "origins pond," Aristotle spawned the most sophisticated view of how organic life may have gotten started. It is clear that the problem of how matter could go from "not alive" to "alive," plagued the Greeks as much as it still plagues mainstream scientists today.

Aristotle's chemistry began with the four elements: earth, air, fire, and water—as well as a fifth, "quintessence," or ether, which he believed existed only in the heavens.

The Animating "Soul"

These elements interacted with one another in each part of the body and were animated by a force he called *pneuma*—best translated in English as "soul."

For Aristotle, living organisms were comprised of a variety of souls that were responsible for growth, motion, sensation, and in human beings, for reason. Because he rejected the idea that the universe and the earth had "a beginning," he was only concerned with explaining how life could form from lifeless matter.

Evolutionary Revelations

Aristotle said: "Animals and plants come into being in earth and in liquid because there is water in the earth, and air in water, and in all air is vital heat so that in a sense all things are full of soul. Therefore living things form quickly whenever this air and vital heat are enclosed in anything. When they are so enclosed, the corporeal liquids being heated, there arises as it were a frothy bubble. Whether what is forming is to be more or less honorable depends on the embracing of the psychical principle; this again depends on the medium in which the generation takes place and the material which is included."

Setting the Stage

The specific living matter to be formed is determined by the interaction of the various elements with their "vital heat," or their pneuma.

Through this view, Aristotle answered the question of how life could come from inorganic matter, laying the groundwork for more modern ideas of "spontaneous generation." On this assumption, it was believed that some organic life (however, explicitly not humans) could spring forth from old rags, grain, or other matter—indeed, even out of thin air.

Spontaneous Life

Aristotle wasn't concerned with "origins," as we are concerned with them today. The spontaneous generation of life was for him a "local explanation" of how organic life might spring forth from an inorganic seedbed of the natural world. It was only later that the question of the origin of life would be entertained and connected with actual chemistry to explain how it might well have all begun.

Following the publication of *The Origin of Species* in 1859, evolution adherents began to speculate about these human beginnings. It should be noted that Darwin (at least in the book's second edition in 1860) stated that life was a product of "creation" rather than of self-generation. And in *Descent of Man*, Darwin goes on to say, "In what manner the mental powers were first developed in the lowest organisms, is as hopeless as how life itself first originated. These are problems for the distant future, if they are ever to be solved by man."

Up from the Mud

Taking a jab at a prevailing theory that life somehow originated from the literal mud of life, Darwin contended that a "mass of mud with matter decaying and undergoing complex chemical changes is a fine hiding place for obscurity of ideas."

In any event, the pressure was on for a viable explanation of how organic matter might have first come from inorganic matter. Whereas some younger Darwinists adhered to "abiogenesis"—spontaneous generation—older ones, such as Darwin's sidekick, T. H. Huxley, were skeptical that this ever happened.

The Miller-Urey Experiment

In 1953, then doctoral student Stanley Miller got a big idea. Having heard a lecture by Nobel laureate chemist Harold C. Urey—in which Urey discussed how the earth's primitive atmosphere would have been fertile ground for molecular formation—Miller created in the lab a "similar" solution rich with hydrogen, water, ammonia, and methane, zapping the gases with electric sparks to simulate lightning.

Within several days of re-creating this "primitive atmosphere," the container was stained with a reddish slime that was rich with amino acids.

> **Evolutionary Revelations**
>
> Even hard-driving evolutionists like Social Darwinist Herbert Spencer believed that simple organisms were too complex to arise from nonliving matter.

Amino acids are compounds that bind to form the proteins which are the fundamental building blocks of life. These exciting results pointed to the possibility that organic life could come from inorganic matter. That amino acids (prerequisite to cellular life) were easy to produce inspired hope that chemistry was on the verge of solving the genesis of life.

Among some chemical problems with the experiment was Miller's introduction of electricity. Used to simulate the lightning believed to have been common on primeval Earth, it is highly unlikely that it would have been as pervasive as Miller's experiment presumed. If so unlikely, amino acids would not have been created in the amounts that Miller had achieved, calling into question a defensible conclusion that life could come from inorganic life.

Just a Matter of Time

Although chemistry has yet to build an analytic bridge between organic and inorganic matter, it has made great strides in explaining organisms in strictly chemical terms.

The vitalism begun by Aristotle almost 2,500 years ago has hung on in various "scientific" forms to the present. Our greater understanding that living processes are by their nature chemical bodes well in the minds of mainstream chemists for the possibility of showing how organic life might primitively form.

For all the experiments that sought to disprove spontaneous generation—however they may have shown maggots and mice don't naturally generate from rags—

abiogenesis was never disproved as an explanation for life, and remains a possibility as chemistry progresses at the molecular level.

And this is the nature of science, mainstream scientists point out. Indeed, they could say that much of the resistance to abiogenesis is in fact motivated by religious and/or philosophical convictions. It is only a matter of time, they might contend, before material processes are unveiled to reveal there is no scientific need to postulate "a designer."

Blinded By Science

Recent simulations conducted at the Universities of Waterloo and Colorado indicate a prebiotic atmosphere as much as 40 percent hydrogen. Such a climate would have been more hospitable to organic life than previously presumed, making for a newborn relevance of the Miller-Urey experiment.

Deoxyribonucleic Acid: We Call It DNA

Until relatively recently, little was known about the life and function of the cell. With dramatic technological advancements and scientific research, we have come to recognize a cellular complexity beyond any prior imagining.

Michael Behe notes that as late as the middle of the nineteenth century, it was supposed—because of limited technology—that cells were a kind of protoplasm. "From the limited view of cells that microscopes provided … a cell was a 'simple little lump of albuminous combination of carbon,' not much different from a piece of microscopic Jell-O," says Behe.

Deoxyribonucleic acid, or DNA, changed all that.

Blueprint of Life

DNA is an acid that contains the genetic instructions which determine biological development. Often referred to as "the blueprint of life," it is cellular material bearing the chemical information needed for organisms to grow and live. Though discovered in the mid nineteenth century, Nobel Prize winners Francis Crick and James Watson are often credited with the discovery of its "structure" (first publishing their findings in 1953 and further developing their discovery throughout the remainder of the decade): a double helix, or two intertwined molecular strands, whose particular design forms "genes."

Genes are parts of the DNA that make the biochemical code, which determines how proteins are created and repaired, to function as the building blocks for life.

From hair texture to eye color to the length of one's legs—even to hand-eye coordination—we are all products of the DNA inherited from our parents. For this reason, athletic scouts are known to keep their eyes on the offspring of basketball stars; indeed, with the growing prospects for genetic engineering, one often hears speculation about the possibility of vending DNA for the sake of buyers' worldly success.

> **Evolutionary Revelations**
>
> The actual practice of DNA science is varied: from criminal investigations to computer science to history to genealogy to anthropological research, as well as to the ongoing studies of the origins of the universe.

If proteins are constructed of amino acids, then how are these acids organized? For proteins to form, they must be arranged in a very specific sequence. As we shall see, it isn't the need for this arrangement that is in question; it is the motivating power that creates the arrangement that drives the ID debate.

Doctor Crick's Hypothesis

Proteins seem too complex to have arisen by chance. By the 1950s, when their vast complexity became increasingly apparent, general scientific laws seemed too general to justify protein construction. Biologists began to research the process, hypothesizing ways that the necessary information could be generated to construct these building blocks of life.

They found it in the coding performed by DNA. Crick discovered that the sequence of amino acids were a function of the arrangement of certain chemicals in the DNA molecule. These "nucleotide bases" located on the spine of the DNA strand are adenine, guanine, cystosine, and thymine—symbolized by the letters *A*, *G*, *C*, and T—and function as the language that informs the building of a protein in a cell.

Crick asserted his "sequence hypothesis," whereby the constitution of amino acids in proteins is a function of the sequence of the bases. Crick and others performed a series of experiments that revealed the informational characteristics of the DNA molecule, and this hypothesis is now considered a part of the "central dogma" of molecular biology. The coding region of the DNA molecule allows for "specified complexity"—that is, not only the potential for complex possibilities in any given protein, but a complexity that is specifically designed to perform certain protein functions.

From Whence Comes the Information?

There are generally three explanations in mainstream science for how this complex and specific information comes about in protein cells. As in physics, random chance has often been employed to answer the mystery. However, in the face of the enormous information that is borne in DNA sequencing, to simply invoke "chance" seems inadequate even to many mainstream scientists.

A second mainstream scientific explanation is prebiotic natural selection. Based on Darwinian ideas, this theory holds that informational complexity is gradually increased by natural selection to the end of more complex cells.

A third proposition is that the information that is found in DNA and in proteins is a product of inherent chemical attractions or other such natural laws. Just as salt and snowflakes naturally crystallize, so DNA might formulate itself. Or as water always tends to order itself as it swirls down a bathtub drain, so might such molecules follow similar laws of thermodynamics.

Fact or Faith _____

Some neo-Darwinists hold to the prebiotic natural selection view. However, it might fail to explain how the information that produced existing cells appeared for selection in the first place.

In any event, to the extent that these hypotheses are still inadequate, mainstream scientists consistently reply that this is part of the process of scientific inquiry. Regardless of how mysterious the natural world appears, it is only a matter of time before the answers will be found.

Rather than resorting to a "God of the gaps" that is invoked in the face of ignorance, mainstream science will finally succeed in rendering a natural explanation.

The Least You Need to Know

- ◆ The ID debate extends to chemistry, where the question of design is now being argued at the molecular level.

- ◆ Carbon is present in all organic life. However, it had yet to be formed at the moment of the big bang.

- ◆ The earth's atmospheres have evolved over billions of years.

- ◆ The first oxygen-producing organisms emerged about 3.3 billion years ago.

- ◆ The discovery of DNA, in its mathematical complexity and perfection, has given rise to the belief that the emergence of life is more than simply a "chance" occurrence.

Chapter 10

Mainstream Biology

In This Chapter

- ◆ A look at biology
- ◆ Clotting blood
- ◆ The amazing eye
- ◆ We're *only* human
- ◆ The all-important fossil record

When we were looking at chemistry, we saw how that field of study is divided between organic and inorganic disciplines; but insofar as biology is concerned with "organisms," all biology is "organic."

In simplest terms, biology is the scientific study of life.

It is the study of things that are—or were—alive, instead of things that never were alive: it's the boy eating cereal rather than the metal spoon shoveling it into his mouth.

Both the boy and the cereal are (or were) living things (and therefore botany is part of the field). Because the only sign of life is growth, biology is occupied with growth, and change, and dying, and how we came to be.

What Is Biology?

Biology concerns itself with all organic life, in its vast diversity of species: from bacteria, to plants, to fish, to reptiles, to mammals, as well as to us human beings.

Biology has recently exploded beyond (indeed, beneath) what the eye can see. With the technological advancements of the last 50 years, research at the particle level has created wholly new scientific fields: molecular biology, chemistry, and genetics, just to name a few. Thus the life of the cell, which until recently could only be conjectured about, has now been revealed as an incredible machine that ingeniously sustains organic life.

As in physics and chemistry, certain general theories govern mainstream scientific thinking. We will see how Darwin's theory of evolution—in a variety of versions—has reigned supreme as a prevailing paradigm by which biology is still conducted. In this context we explore the issues most relevant to the Intelligent Design debate: namely, creation of body parts, the development of species, and the evolutionary evidence of the fossil record.

The Complexity of Biology

Though we will reserve a more detailed discussion of "evolution" until a later chapter, it is important to understand that modern biology is so steeped in evolution and evolution's assumptions that it is impossible to fully avoid it now.

Here's why: although mainstream biology clearly acknowledges the seemingly miraculous complexity of organisms, it takes issue with ID's belief that any of it had to be "designed." Darwinian evolution seems so startlingly perfect as a self-sufficient explanation that the introduction of some vague and unseen designer appears, at best, unnecessary.

Such scientists are not impressed with the idea that today's scientific mystery is anything but tomorrow's bridge to an evolutionary explanation. So they are inspired to reveal the evolutionary pathways that lead to the creation of complex organisms. To the extent that any lack of such evolutionary pathways still plagues our current understanding, these scientists believe it's just a matter of time before Darwin will be vindicated.

What is missing for most mainstream biologists in the argument for design is the idea that spare parts in lower organisms can be used to build more complex structures. The pathway of gradual "complexification"—with regard to any organism—has become the necessary proof text for the authenticity of evolutionary theory. Therefore significant effort has been made to show how this creative process might work: with regard to such things as the bacterial flagellum, the blood-clotting system, and the eye.

The Little Engine That Could

Evolutionary biologist Kenneth Miller has become a leading critic of Intelligent Design. Miller believes that ID is based upon a false assumption that random mutation and natural selection cannot create complexity. He calls this an "argument from personal incredulity," in which organic complexity is used as a weapon to convince a naïve public of the need for life to be "designed."

So Miller points to the bacterial flagellum as a culprit in the ID agenda. He notes that the flagellum has been used so often it has become ID's "poster child," whose underlying program is to try to undermine evolutionary theory.

The bacterial flagellum is like a tiny outboard motor with myriad moving parts: most notably, a rotary hook that drives a filament (think of fishing line), both of which are driven by its engine. It was discovered that certain bacteria use the flagellum to swim—which boasts an ingenious acid-fueled energy generation machine. Behe's belief that the bacterial flagellum is irreducibly complex means that it can't be broken down into less complicated parts created by evolution.

But Miller asserts that recent research has revealed its reducibility. He explains that as a result of scien-tific progress in related gene and protein research, we now know that there are indeed precursors—forerunners—to the flagellum. Therefore Behe's argument, Miller contends, contradicts its own prerequisite: that for an organism to be irreducibly complex, there can't be evolutionary pathways.

The type three secretory system (TTSS), which enables bacteria to inject toxins through the membranes of cells, exhibits similarities with the bacterial flagellum at the molecular

Fact or Faith _____

Biochemist Michael Behe's contention (which we will spend some time with in Chapter 13) that the flagellum can't have evolved, is what Miller and other evolutionary biologists try to address.

Blinded By Science _____

Stated directly, the TTSS does its dirty work using a handful of proteins from the base of the flagellum. From the evolutionary point of view, this relationship is hardly surprising. In fact, it's to be expected that the opportunism of evolutionary processes would mix and match proteins to produce new and novel functions.

—Kenneth Miller, _The Flagellum Unspun_

level. Insofar as they share proteins from a common origin, some mainstream biologists contend that these two molecular machines are related by evolution. The so-called "irreducible complexity" of the bacterial flagellum is, in Miller's mind, a clear example of gradual development.

The Blood-Clotting Factor

Let's look at the biology of the blood clotting system and why it's an important ID issue. As an example of irreducible complexity, it is used to invoke design as the most likely explanation for how certain biological features came to be. It is therefore important to understand this amazing, complex process to see what is at issue for mainstream biologists.

Ken Miller believes the human clotting system is in fact "remarkably simple." He explains how "fibrinogen," a fibrous protein solution that makes the clots, has a sticky part at the molecule's center, bathed by chains of amino acids. These chains, which all bear a negative charge, keep the molecules apart; but when a clot forms, the chains are clipped by a protein-cutting enzyme, thrombin. With the sticky parts of the molecules exposed, the fibrinogens stick to each other, and so, he says, the basic blood-clotting process involves just two molecules.

The X Factor

Like fibrinogen, most of the time thrombin has to stay inactive; otherwise, our circulatory systems would be one big clot, and we would die. But its inactive form, prothrombin, must be switched on to start the clotting process. Prothrombin is turned on by "factor X," which clips off its inactive protein in order to produce the thrombin that activates clot formation.

So what activates factor X to begin the clotting process?

Several more catalytic factors interact to create a "cascade," which results in the scab on a young boy's knee that literally saves his life. This cascade process, Miller explains, amplifies the original signal so that a single active molecule is magnified more than a million times. With numerous "falls" in the clotting cascade, the clot can form more quickly, while minor injuries, giving lesser signals, still get their due attention.

The Sticky White Blood Cells

In any case, Miller believes Intelligent Design's contention that the blood-clotting system is "irreducibly complex" is simply incorrect. As an evolutionist, he asserts the cascade was created gradually—that it was built from existing parts and was able to clot in more primitive forms. Even starting 600 million years ago in small prevertebrates, sticky white blood cells could have done the job in a more primitive way.

The Hyper-Advanced Human Eye

"What good is the use of half an eye?" neo-Darwinist Richard Dawkins asks. Of course, Dawkins is setting up a "straw man" in order to burn it on the spot.

> **Evolutionary Revelations**
>
> "What good is the use of half an eye?" Actually, this is a light-weight question, a doddle to answer. Half an eye is just one percent better than 49 percent of an eye, which is already better than 48 percent, and the difference is significant.
>
> —Richard Dawkins, "Where'd You Get Them Peepers?" (*New Statesman & Society*, June 16, 1995)

In any case, this is an effective way to catalyze lively discussion between mainstream science and Intelligent Design about the logic of evolution.

Dawkins is responding to the argument launched by Intelligent Design that there isn't enough time in geological history to have evolved a human eye. His argument assumes that the vast complexity inherent in the eye could have been produced by evolution—indeed, must have been evolved. Unlike proponents of ID who are stymied by this vast complexity, Dawkins and his allies tend to shrug their shoulders, amused by ID's "incredulity."

Suffice to say that the structural complexity of the human eye makes the bacterial flagellum look as simple as a kitchen drain. Selecting a single photocell as one part of the human eye, Dawkins points out that there are myriad other components of this visual system. Composed of a nucleus, cellular layers to catch protons, and connecting wire, the photocell contains mitochondria, which are themselves, enormously complex.

"Mitochondria are found not just in photocells, but in most other cells," writes Richard Dawkins in *The Blind Watchmaker*. "Each one can be thought of as a chemical factory which, in the course of delivering its primary product of useable energy, processes more than 700 different chemical substances, in long, interweaving assembly-lines strung out along the surface of its intricately folded internal membranes."

Dawkins adds fuel to evolution's fire by underscoring that this intricate "design" is repeated 125 million times in each retina.

Dawkins reassures his readers that the eye is no problem for evolution: in fact, two Swedish researchers have even shown how long the eye would take to evolve. Dan Nilsson and Susanne Pelger used computer simulation to calculate that it would take less than half a million years for a functional eye to evolve. Dawkins says, "In the light of Nilsson and Pelger's results, it is no wonder 'the' eye has evolved at least 40 times independently around the animal kingdom."

So for Dawkins, the mainstream argument holds: the eye could have easily evolved. Neither time nor complexity is a problem for Darwinian evolution. The argument for design is an unnecessary argument from "incredulity."

Fact or Faith

Richard Dawkins says that an eye creates a photographic resolution of about 5,000 times the resolution points that constitute a good magazine photograph. Although Dawkins may admit to its complexity, the eye, he says, is simply not "designed."

Special to Be Human

If biology is the study of life, and those studying it are human, it is only natural that what is focused on is the study of human life. And indeed, in the light of Intelligent Design, the origins of human life have become a critical centerpiece of the whole scientific debate. We deal in greater depth with this question when we come to the chapter on evolution; but for now we look at the biology that lays the general groundwork.

As we've already noted, it is difficult to separate biology from "evolution." For many, evolution *is* biology—at least in its most popular form: descent with modification by random mutation and natural selection. The truth is there is much heated debate about the meaning of the word *evolution*, and there are those who contend that neo-Darwinists completely misunderstand Darwin.

In any event, we'll look at the creation of species from a mainstream biological perspective. Macroevolution is a critical player in the human scenario, leading to the ultimate mainstream conclusion that we all descend from a common ancestor. But as we shall see, these assumptions will be called into question by Intelligent Design—remembering, however, that there is also disagreement within each scientific camp.

Odd Links in a Great Chain

"Speciation" is the evolutionary process whereby new species are created. This is important because species are the links in the biological "great chain" of being. There are several different ways that species arise, but one common characteristic is that speciation comes about by the condition of geographical isolation.

Every species has a habitat that sustains and nurtures it; when a population migrates, or is somehow physically separated from its original habitat, it eats different kinds of food and is subject to different environmental influences. In the process, it undergoes genetic change as a result of different "selective pressures." So over huge tracts of time and change, if these separated species were to come back into contact with each other, they couldn't reproduce: the defining mark of the conception of a new species. Indeed, this is natural selection at work—descent with modification—creating new species, and the diversity of life that we all see every day.

There is much debate about the speed with which new species are created in the chain. When we look at the *fossil* record, we shall see that evolutionists disagree about the frequency with which speciation happens (as well as its significance). What isn't in conflict for mainstream biologists, is that evolution spawns new organisms—accounting for the genetic differences between algae, ants, apes, and human beings.

def•i•ni•tion

> **Fossils** are the traces or remains of formerly living organisms: including animals, plants, fungi, bacteria, and other life forms that once existed. They are now preserved in the earth's crust, as well as in rocks and other above ground fragments. Fossils can appear anywhere on Earth—the locations are hard to predict. And as many locations as there are, the possibilities for their discovery are almost endless; hence, the criticism that the record is incomplete says only that the work is ongoing.

Macroevolution vs. Microevolution

The difference between macroevolution and microevolution is first and foremost, a matter of size. But if "macro" means big and "micro" means small, where is the dividing line? The answer is that the level of species is the boundary between the two—having serious implications for their social lives, because as we've said, they can't reproduce.

If evolution means change, then microevolution means change within a species; whereas macroevolution means change between or beyond a particular species. There are numerous theories about how this change occurs, but the popular "modern synthesis" asserts that macroevolution is an accumulation of microevolutionary events. When ID critics say they believe in microevolution, but not in macroevolution, mainstream biologists respond it's just a matter of degree and that the same genetic principles hold.

Thus, for mainstream biologists, this ID criticism looks like religious rationalization. They might argue it's a holdover from creationist thinking that human beings are "set apart" by God, and that mingling their genes with lower animals defies "divine providence."

Nevertheless, according to the mainstream biological perspective, this isn't good science; whatever one's beliefs, one can't argue with evolution's evidence.

In the article, "Macroevolution," (Talkorigins.com, 1997), John Wilkins says:

> There is no difference between macro- and microevolution except that genes between species usually diverge, while genes within species usually combine. The same processes that cause within-species evolution are responsible for above-species evolution, except that the processes that cause speciation to include things that cannot happen to lesser groups, such as the evolution of different sexual apparatus (because, by definition, once organisms cannot interbreed, they are different species). The idea that the origin of higher taxa, such as genera (canines versus felines, for example), requires something special is based on the misunderstanding of the way in which new phyla (lineages) arise. The two species that are the origin of dogs and cats probably differed very little from their common ancestral species and each other. But once they were reproductively isolated from each other, they evolved more-and-more differences that they shared but the other lineages did not. This is true of all lineages back to the first eukaryotic (nuclear) cell.

Universal Common Descent

Universal common descent is an evolutionary term that refers to the idea that all living organisms descend from a common ancestor. We are connected by virtue of the evolutionary fact that we come from a common "gene pool."

It is an assumption of mainstream biology that the empirical evidence for common descent is simply overwhelming.

Until the advent of molecular biology, we could only hypothesize that human beings came from the same ancestor as the chimpanzee. In conjunction with the fossil record, says neo-Darwinist Richard Dawkins, recent genetic research tells a different story from the one told in Genesis.

Fact or Faith _____

The basic premise of universal common descent is that, not only are we all related by our lineage to Adam and Eve, but that we also are organically related to the flora and fauna that graced the Garden of Eden.

"The last common ancestor of humans and chimps lived perhaps as recently as five million years ago, definitely more recently than the common ancestor of chimps and orangutans, and perhaps 30 million years more recently than the common ancestor of chimps and monkeys," writes Dawkins in *The Blind Watchmaker*. "Chimpanzees and we share more than 99 percent of our genes."

The marriage of Darwinian evolution and newborn genetic research has strengthened the mainstream scientific case for the "fact" of common descent. Now it is felt that there is concrete evidence at the molecular level. The difference between macro- and microevolution is simply a matter of degree; in any case, for these biologists, evolution clearly takes place.

A Holy Grail?

For many biologists, the fossil record is the "Holy Grail" of evolution. It is the material evidence at the scene of the crime; it is the answer to "Where's the beef?" It is the window into history that scientists need—because after all, they weren't around.

Like all "Holy Grails," the fossil record is at the heart of the ID dispute. Mainstream biologists use it to show evolution at work—in petrified clay. ID biologists claim the opposite: that the fossil record is proof that evolution is an inadequate explanatory theory.

Once again, it is important to note that within each respective camp is heated disagreement about the fossil record's importance, as well as its significance. We shall see how some proponents of Intelligent Design believe that it is critical, whereas others (including Michael Behe) treat it as irrelevant. With regard to mainstream biologists who assent to its significance, there is still dispute about the so-called gaps present in the fossil record.

Mortar for the Gaps

A pervasive criticism of evolution by proponents of Intelligent Design is the myriad "gaps" in the fossil record which call the theory into question. Mainstream biologists, especially those who work in *paleontology*, explain this is the nature of all scientific inquiry, and nothing to be concerned about. As work progresses on the fossil record, these "gaps" are being constantly filled in; and in some cases, have even served to vindicate the predictive power of evolution.

def•i•ni•tion

When most people think of **paleontology**, they think of the study of dinosaurs. But it is much more than that. According to Cornell University's Paleontological Research Institution, paleontology is "the study of the history of life on Earth, as reflected in the fossil record."

For example, because evidence pointed to the first mammals being land-dwelling, four-legged creatures, one might think that there would have been four-legged ancestral whales in the fossil record. Indeed, creationists saw the absence of such fossils as evidence against evolution, and in favor of whales being created as a whole and divinely fit for their oceanic habitat.

Yet "New discoveries from the past decade have closed the terrestrial mammal-whale gap in spectacular fashion," writes John Stear in *A Critical Look at Creationist Paleontology*. "Several very early whale specimens are now known, and these convincingly illustrate the relationship of early whales to their terrestrial mammalian ancestors. While no modern (whales) have legs, the earliest whales all apparently possessed legs, losing them gradually over a period of about 10 million years."

Fossils Do Tell Tales

Thanks to the fossil record, we can embrace the breadth of our organic history. We know that the first multicellular animals appeared 600 million years ago; we know that marine vertebrates, derived from invertebrates, evolved to become reptiles, mammals, and then humans. Indeed, we know about the dinosaur age, which may have ended by a meteor, allowing mammalian life, by this fantastic accident, to evolve into our modern human form.

Harvard paleontologist, Stephen Jay Gould, warns against a "progressive" view in which it is implied that all this evolution happened for the sake of humans. It is a tenet of Darwinian evolution that selection is "nonprogressive;" that is to say, that change occurs by way of fitness to environments, and not by fitness to some abstract "purpose." In any case, in the mainstream biological view, the fossil record—"gaps" and all—is consistent confirmation of natural selection, producing gradual descent with modification.

Getting It Right

Perhaps the first thing most mainstream biologists would say is that the fossil record is not perfect. For several reasons it is considered incomplete as a history of organic life—not the least of which is it literally happens in the geological mud and muck of life. When one understands the fossilization process, as well as the record's quirky nature, one appreciates the reason the fossil record may appear to be riddled with gaps and holes.

It is significant to note that the fossil record is prejudiced toward hard-shelled organisms. Soft-shelled organisms are underrepresented, and for good reason. They tend to decay before they fossilize—as do organisms living in areas such as rain forests where less fossilization occurs—and because of the fossilization process, physical distortion can occur, making it difficult to reconstruct the place of soft-shelled organisms within the record.

The point is that like all scientific research, paleontology is imperfect—and so it is unfair to expect perfection when perfection is impossible. Still, the fossil record conveys a history that is invaluable to our understanding; and the evidence, for most biologists, points to an evolutionary process. Yet we shall see how even within the evolutionary camp, there is significant conflict—specifically, by those who interpret the "gaps" in terms of "punctuated equilibrium."

"Punk Eek"

To give you an idea of just how fiery the conflict is within mainstream science, one need only look as far as Stephen Jay Gould's theory of "punctuated equilibrium."

With Niles Eldridge, Gould developed a theory that the absence of transitional fossils—that is, fossils that embody "intermediate forms" between extreme forms thought to be connected—was due to the fact that these apparent jumps between species in the fossil record did not represent incomplete information, but real, sudden, punctuated change. This was so much a departure from Darwin's contention that change occurred gradually, that more orthodox Darwinists not so generously called the theory, "evolution by jerks."

But Gould and Eldridge felt that ongoing change was not the normal state of life. If organisms underwent physical change, most of the time it was minor. The status quo of evolution was equilibrium, or stasis. When large changes occurred, Gould contended, they occurred in small populations where change could have a significant impact on genetic reproduction. This "punctuated" change happened so rapidly, he said, that there was little time to leave transitional forms, thereby explaining the gaps in the fossil record.

It is not a matter that fossil record critics do not believe in evolution. It is rather a matter of the best interpretation for how organisms evolve—in contrast to proponents of ID, who believe this calls evolution into question. Yet adherents to "punk eek" see evidence of the theory in the fossil record: in a marine micro-fossil, in the trilobite, and in the beloved Tyrannosaurus Rex.

The Least You Need to Know

- Biology concerns itself with all organic life, in its vast diversity of species.

- Modern-day biology is steeped in evolution and evolution's assumptions.

- Leading ID proponents believe that the "bacterial flagellum" (the little engine that could) is so irreducibly complex that it cannot be broken down into less-complicated parts created by evolution.

- ID proponents point to blood clotting and the complexity of the human eye as proof of intelligent design.

- Speciation is the evolutionary process whereby new species are created. This is important because species are the links in the biological "great chain" of being.

- For many biologists, the fossil record is the "Holy Grail" of evolution.

Part 4

Intelligent Design Applications

Of course ID proponents recognize the "big three" scientific disciplines. In many ways it embraces them, but not without criticism regarding some of the "questions" it believes go unanswered by mainstream science.

Here we look at why ID proponents believe there must be a "designer" or, for most, a God behind—or beyond—the physical world.

WHEELER '06

Chapter 11

Intelligent Design Physics

In This Chapter

- ◆ A big bang argument
- ◆ A way to find God
- ◆ Heat and work
- ◆ The critical question

If there is one scientific discipline that seems naturally suited to the concept or tenets of Intelligent Design, it is probably physics—with its focus on creation, its cosmological laws, and its "finely tuned" universe.

After all, even those who don't believe in—or reject the idea of—an "Intelligent Creator" have to work their way around the apparent fine-tuning of the universe. And arguing around that is not easy. Unlike in biology or chemistry, where ID has generally been on the defensive, in physics proponents of Intelligent Design may enjoy a bit more confidence.

The appearance of cosmological design seems to naturally beg the question: how could myriad factors so precisely conspire to (if we may be so bold) produce "life"? For theists, the answer is simple enough: God created the universe just so. The purpose was God's from the beginning of time, and we are here to enjoy it.

But for atheists, for whom design is ultimately seen as some kind of physical illusion, life's "apparent" hospitality is generally perceived as a product of happenstance. In other words, there is no purpose in it. As they say, stuff just happens.

In any case, the advent of the big bang theory was a boon to Intelligent Design. Moreover, recent research, revealing the delicate balance of the universe, is further indication to ID proponents of the existence of a designer. To some of them, the second law of thermodynamics reveals that there has to be a countervailing force that holds the world together when it really wants to fall apart.

Big Bang, Big Defense

As we have discussed, Intelligent Design goes back to a time before the ancient Greeks. And since the time of Copernicus, science continued to recognize a designer as a common assumption that needn't conflict with a scientific point of view. Yet as the power of science has risen to challenge religious understanding, the design perspective has had to appeal to scientific expressions of design.

As we know, big bang theorists estimate that the universe was born some 13 billion years ago, in a super-hot, fiery explosion which began the whole thing. The ongoing expansion of the universe, which the big bang theory predicted, has since been observed by "red shift" light that indicates receding galaxies.

High-powered telescopes that were unavailable when the big bang theory was conceived have gathered evidence, which for some scientists, points to the possibility of God.

George Smoot's "Wrinkles"

In 1992, astrophysicist George Smoot, using the Cosmic Background Explorer satellite, found evidence from the early universe he saw as seeds of the galaxies. Until then, the smooth background radiation that served as evidence of a big bang seemed to contradict the lumpy universe that was observed in the observable heavens. With the data he collected, Smoot discovered inconsistencies that looked like wrinkles that pointed to our cosmic beginnings, and the universe's lumpiness.

Albert Einstein, whose relativity theory substantiated the big bang, resisted its theistic implications before finally admitting a "beginning." You may remember that in the prevailing "steady-state" theory of the universe, space and time were like a repository in which physical elements interacted. But with startling evidence that all we know—

time, space, matter, physical laws—had one distinct beginning, one had to ask: what of the beginner?

Evolutionary Revelations

In a crowd of reporters, Smoot was heard to say, "If you're religious, it's like seeing God." (Dick Teresi tells how several weeks after this remark, graffiti was found in Smoot's laboratory hallway that said, "If you're God, it's like seeing George Smoot.")

The Argument Begun

Astronomer Hugh Ross, who wrote *Creator and the Cosmos*, is a creationist voice in the current debate over Intelligent Design. To Ross, as to many ID proponents, the big bang's implication is simple and basically inescapable (if not historically familiar): "If the universe arose out of a big bang, it must have had a beginning. And if it had a beginning, it has no choice but to have a beginner."

Fact or Faith

Hugh Ross's argument basically goes like this: everything is caused by something other than itself. Therefore, the universe in which we find ourselves was caused by something other than itself.

This may be familiar to those who remember Thomas Aquinas's style of thinking. Aquinas systematically constructed "five ways" for the proof of God's existence—the last of which hinged on the apparent design of the universe. Ross, however, is here employing Aquinas's "second way": the argument from "efficient causes," in which the nature of our causal universe leads us to know that God exists.

Aquinas goes on to say that the chain of causes cannot be infinitely long. There has to be a first cause to which all subsequent things in the universe go back. For Aquinas, the name for this "thing" is God, as it is for many ID proponents.

The Argument Evolved

Although the big bang argument (for the existence of God) shares some of Aquinas's assumptions, it differs in its need to recognize the bang as the start of all causality. Remember this theory also says that chaos reigned before the universe was created—when all laws, including causality—also came into being. So ID must qualify that everything that has a beginning also has a cause. Such laws as causality have no meaning until the beginning of time.

But the universe in time is what we're talking about.

Fact or Faith

Not only does the universe have to have a cause, according to Hugh Ross's reasoning. That cause, also according to Ross, is nothing less than "God."

This big bang argument improves on Aquinas's in an interesting way. Aquinas's contention is that everything in the universe has to have a cause. But this means that God has to have a cause, which undermines the definition of God as one who, by his preeminent nature, is necessarily uncaused.

Relativity asserts there was no time before the universe's creation. So causality, as a dynamic that operates within the universe, need not operate before the universe, and therefore God need not have a cause. Thus in the big bang model, God can retain the position of "the uncaused cause," residing in a realm where such laws do not apply, and where God's existence still can be asserted.

Confused? It's simple: God, by virtue of being God, was—according to relativity and causality—able to exist before the big bang.

Finding God by "Abductive Inference"

A question that one often hears in arguments between mainstream science and Intelligent Design is how ID can contend it is able to find God in the material universe. Doesn't God transcend or go beyond the universe, mainstream science might appropriately ask? How is it that one could possibly "prove God" within the physical universe?

For Intelligent Design, the answer may lie in the term *abductive inference*. By making suppositions from experience, which connect cause to effect, ID advocates believe that they can show how God is "the best explanation." In the same way that black holes were inferred 30 years before they were physically observed, God can be inferred by the manifest design of the physical universe.

The Big Three

Charles Thaxton, a leading voice in ID, contends a third inferential method needs to be employed in scientific inquiry—alongside deduction and induction. Deduction (you might remember from high school science class) moves from the general to the particular: the sun is shining, and therefore I can see the threads of the baseball I am holding in my hand. Induction, on the other hand, moves from the particular back to the general: I can see the threads on the baseball in my hand, so therefore, the sun must be shining.

Abductive inference (if one has a third hand handy) reasons back from experience to cause. Because it avoids drawing deductive conclusions that are narrowed by prior assumptions, and because it is open to possible causes beyond physically restricted induction, abductive inference is able to entertain physical possibilities that these other two reasoning methods cannot in the process of doing science.

Although this may seem far-flung from the mainstream approaches of scientific inquiry, such an approach to "intelligent causes" is currently used in science.

Evolutionary Revelations

… deep channels or rills on the Martian surface are so similar to what we know by experience to be the result of running water, that we would associate the natural cause of the channels with water—even if there is no water on Mars today. Thus scientists at NASA have concluded that water must have been on Mars sometime in the past. On the other hand, were we to hike in the black hills of South Dakota and come upon granite cliffs bearing the likenesses of four United States Presidents, we would likely identify Mount Rushmore as the work of artisans instead of a product of wind and erosion. Our accumulated reservoir of experience enables us to discriminate types of effects we see and to distinguish a natural process from an intelligent cause.

—Charles Thaxton, "A New Design Argument," *Cosmic Pursuit,* March 1, 1998

NASA's Search for Aliens

The U.S. government is clearly invested in abductive inferential research, funding the "Search for Extra-Terrestrial Life" program (SETI, for short). Scanning the heavens for artificial electromagnetic impulses, NASA has predicated its work upon potential "intelligent" causes. The complaint from modern science that "intelligent causes" are beyond the domain of science is, for advocates of Intelligent Design, therefore simply untrue. Of course, as with any philosophical argument, there are always counterarguments. In the case of mainstream science, one might be that "divine intelligence" cannot be tested, whereas the existence of extraterrestrial life could always be falsified.

In any case, it seems that the question of design is at issue precisely because, whether or not life is in truth designed, it *appears* to most of us to be designed. The mystery is its "source." For some it may be chance, while for others purely natural causes such as self-organizing mechanisms that preclude the necessity of God. We have seen how in the mainstream scientific fields of chemistry and biology that much has been made of "self-organization"—instead of any outside "agent." Intelligent Design proponents argue such self-organization is unscientific; specifically, insofar as it denies the second law of thermodynamics.

The Second Law of Thermodynamics

The first law of thermodynamics states that energy is always conserved; the second law describes the relationship between "heat" and "work" (in Greek, *thermo* and *dynamic*). The first law is the reason that no one has invented a perpetual motion machine, and the second law explains how, by concerted effort, work produces a tidy bedroom. Indeed, the second law is expressed in terms of "entropy"—the measure of disorder in a system—as heat flows from a warmer body to a cooler one, entropy, disorder, increases.

Which is to say that heat represents an order that organizes the potential for "work." Left alone, a system will always move from relative order to disorder, thereby increasing a system's inherently entropic tendencies. And left alone, the temperatures in any given system tend toward evening out—so the heater in my bedroom runs out of fuel, and the net productive work decreases.

The Bad News

The bad news is that entropy is the way of the universe. Disorder, chaos, and randomness all conspire to disorganize the world. Thus we have inherited a chaotic world, which could lead one to believe it has no purpose. And isn't one of the features of God to impose a divine order on creation? How could God create a world where chaos reigns; doesn't this question God's existence? Or if God does exist, doesn't this imply that his creation lies beyond his own control? (Therefore, what does it matter if there is a God, if he is powerless within his universe?)

One of the contentions of Intelligent Design is that since nature goes from order to disorder, no system can spontaneously organize itself without the help of some outside force. In other words, unless I go into my room and clean it up myself, the dirty laundry has little chance of finding its way up off the floor and out of the room—let alone, down the hallway, into the laundry room, and finally the washing machine. And if I left my room for good, and the heater that had heated the room ran out of fuel, entropy would have continued to increase toward ultimate deterioration.

The Good News

If entropy is the way of the world and, indeed, of the universe, then how is it that order can be introduced into any material system? Shouldn't the world be an irreversible mess in which no work could ever be accomplished? How can we explain the purposeful direction apparent in human life?

These are the questions that are asked by ID in its quest for the "agency" that brings order out of chaos, that organizes human life, and that gives meaning to the universe.

The answer is the universe appears to be created and sustained by an intelligence, which in some way intervenes in the natural world to create its apparent order. Although for most ID scientists its identity cannot be scientifically discerned, they are willing to assert that transcendent agency is the most successful explanation. If the big bang, as a singular proof of God's existence, started the whole thing—and entropy, as a degenerate force, must be reversed by some outside intervention—then the apparent fine-tuning of the universe, with its seemingly miraculous laws, provides ongoing testimony to the simple fact that the universe has been designed.

> **Evolutionary Revelations**
>
> In the field of biology, as we shall see, this argument is critical: how the seeming miracle of organic life could come from random processes. In this way, the second law of thermodynamics—a fundamental law of physics—bears on the rest of the ID argument in its conflict with mainstream science.

A Scientific Postscript

It should be noted that there are ID advocates who don't subscribe to the argument that the second law of thermodynamics means that all things move toward disorder. This idea stems from creationist thought—particularly, that of Henry Morris. To some ID minds, it isn't relevant; to some, it is simply wrong.

The idea that entropy inevitably increases only holds for the overall system, and not for interacting subsystems or for the system's surroundings. The heat flowing out from my heaterless room increases the room's entropy, while the heat that dissipates into the room's surroundings decreases the surroundings' entropy. So an increase in order need not result from an outside "intelligent agent;" it can happen as a "passive" consequence of heat being infused into another context.

Thus it should be noted that there are clear examples of order coming out of disorder. Snowflakes spontaneously form from random, moving water particles, and salt crystallizes form by the evaporation of water from liquid solutions. Yet even these examples, ID scientists might say, smack of predetermined design; so the argument lies in the reading of the science rather than the findings themselves.

A Fine-Tuned Universe

Though many scientists agree—and will indeed admit—that at least the universe *seems* designed. The question of whether or not it *is* designed remains the critical one.

In his book *The Emperor's New Mind*, mathematical physicist Roger Penrose calculates the accuracy with which the universe's constants had to be chosen. It is on the order of 10 raised to the power of 10, raised to the power of 123. If one wrote out the number, with each zero miniaturized to the size of a proton, this definition of the necessary accuracy wouldn't fit in the known universe.

Fact or Faith

Mainstream scientists often suspect a nonscientific, religious agenda, whereas ID scientists appeal to the evident improbability of the universe.

According to Penrose, "Such figures are beyond human comprehension. We cannot help but wonder how these constants could have happened to have these values. Just a slight change, in even a few of these numbers, and atoms could not form. Stars could not shine. The earth and its inhabitants could not even exist. The universe would just be an enormous junkyard of useless, practically inert particles, dead matter, and chaos."

"Ready! Aim! Fire!"

Suppose you were brought before a firing squad of no less than 50 marksmen. Blindfolded, you wait. Then 50 rifles fire. When the smoke clears, there you are still standing. As the captain unmasks you, you begin to realize that not only are you still alive, but not a single bullet even grazed you.

What do you suppose? Do you believe there was design in this? Or do you believe that you survived by accident?

The firing squad analogy has been often used in the Intelligent Design debate. The odds against random chance being the reason for having been saved from death by a firing squad are equivalent to the odds against random chance producing the universe.

In both cases we are living by reason of design—whether a merciful firing squad or, perhaps, God—so for ID proponents, the mainstream explanation of random chance simply doesn't hold water.

Here to Tell About It

Unlike mainstream science, Intelligent Design tends to place human beings at the center. This is one of the criticisms launched by evolutionists against an "intelligent creation." In physics, this self-centeredness is similarly addressed by the weak anthropic principle—which asserts that all this principle finally describes is what we already knew: that we are here.

So weak anthropic principle advocates might say there was no "miracle" in surviving the firing squad, because the million times you wouldn't have survived, you wouldn't be around to tell about it! But believers in Intelligent Design respond that, in fact, they *are* around; and these "ifs" that are proposed are examples of bad science, and depart from the material evidence. Because we are around to observe and tell about it, we need to examine this fact—as uniquely gifted beings, whose self-conscious understanding is evidence of some kind of design.

Chance Product of a Multiverse?

We know one of the ways mainstream science gets around the strong anthropic principle is to say there is a multitude of chances for survival—or possibilities for a universe. In other words, if there were a million universes, each with its own unique laws, then ours is simply one among these million universes that happened to survive to tell about it. So there's nothing special in our having our own laws—after all, we had to have *some* laws—and any "fine-tuning" is no more than "random tuning" that happened to strike our particular chord.

The trouble is that there is no such evidence of any other universe. So advocates of Intelligent Design may be inclined to scratch their heads about the mainstream contention that it is ID, which is inherently "unscientific."

If God transcends the physical, then so do universes for which there is no concrete evidence—leaving the debate between these two competing camps on an oddly metaphysical level.

Evolutionary Revelations

If we suppose that the universe came about by chance, we must cling to the preposterous idea that our universe beat these overwhelming odds just by some freak accident. Surely no reasonable person would believe such a thing to be true. The only feasible explanation that can account for the improbable, incredible delicate balance in nature is that an immense intelligence designed it for the express purpose of enabling life to exist. Our universe shows such fine-tuning that it is evidently the result of careful and meticulous planning. At least, that is the most obvious thing to conclude. One must recognize that this is not a strict proof of the existence of a creator, but is rather a demonstration that it is the most reasonable conclusion to draw.

—Roger Penrose, as quoted by Michael J. Hurben in his article, "On Universes and Firing Squads or 'How I Learned to Stop Worrying About the Origin of the Cosmos'"

Mainstream science is suspicious (and rightfully so) of specific religious agendas. To the extent that this complaint attacks more than a "straw man," and defends good scientific inquiry, advocates of Intelligent Design might agree with this mainstream criticism.

In fact, here is a frequent point of departure between ID and creationism: the insistence by the latter of naming a specific creator as divine architect. Mathematical physicist Roger Penrose is one such scientist who entertains the idea there is Intelligent Design, yet refuses to suggest a creator.

The Least You Need to Know

♦ Physics may be the scientific discipline most naturally suited to the concepts or tenets of Intelligent Design.

♦ Astrophysicist George Smoot found "wrinkles" in the early universe that are believed to be the seeds of galaxies.

♦ Astronomer Hugh Ross believes that everything is caused by something other than itself, thus the cause of the universe must be God.

♦ ID proponent Charles Thaxton says abductive inference (such as NASA has used in its "listening" for extraterrestrial impulses from space) should—like deductive and inductive reasoning—be employed in scientific inquiry.

♦ The firing squad analogy is often used in the Intelligent Design debate.

Chapter 12

Intelligent Design Chemistry

In This Chapter

- ◆ Design in chemistry?
- ◆ Answers to the earliest questions
- ◆ A courier from God

As discussed in Chapter 9, chemistry is the study of the structure and properties of physical matter, focusing on how the various physical things interact with one another.

Because chemistry deals with tiny things the naked eye can't see, and technology has allowed us to see it at the molecular level; the field has become increasingly abstract—a far cry from the high school lab, with its bunsen burners and beakers of boiling chemical solutions. When we look at some historical ideas of chemical elements—from life arising from mud, to alchemy which promised to turn lead into gold—we are struck by how much we have come to know in the short course of 100 years, having cast off supernatural ideas removed from the modern truth.

We have seen how mainstream chemistry presumes an "autonomy": a self-organizing infrastructure that allows the elements to act on their own. Whatever chemical design may appear to the eye is, in the end, only "apparent." To the extent that there appears to be physical design, it is the product of chance, or evolution—an assumption we first saw applied in physics, and our cosmological beginnings.

So what about Intelligent Design's approach to chemistry?

ID and Chemistry

Not unlike the field of physics, proponents of ID are unabashed about their belief in design in the field of chemistry. In the same way that "fine-tuning" has been applied to the universe by physics, we will see how it is similarly applied to the chemical elements of nature.

We will also see how, rather than taking pains to entertain "abiogenesis," ID asserts that going from inorganic to organic life is impossible. And finally, DNA, the code of human life, is perceived by ID advocates as a quintessential sign of natural design, and for most, the very signature of God.

Chemical Fine-Tuning

One of the unwritten precepts underlying many ID arguments is that this fine-tuned universe in which we live is fine-tuned for the sake of human beings. The question of teleology, or purpose—of why the universe is fine-tuned—is, for the majority of mainstream scientists, "the elephant in the room."

ID proponent and geneticist Michael Denton asserts that this *anthropocentric* perspective is to be admired.

def•i•ni•tion

Increasingly used from the beginning of the twentieth century, the adjective **anthropocentric** simply means man-centered or human-centered. It suggests that man has the greatest value of all things in the universe, and that the universe and everything in it were created for human beings.

In *Nature's Destiny*, Denton writes:

> The anthropocentric vision of medieval Christianity is one of the most extraordinary—perhaps the most extraordinary—of all the presumptions of humankind. It is the ultimate theory and in a very real sense, the ultimate conceit. No other theory or concept ever imagined by man can equal in boldness and audacity this great claim—that everything revolves around human

existence—that all the starry heavens, that every species of life, that every characteristic of reality exists for mankind and mankind alone. It is simply the most daring idea ever proposed. But most remarkably, given its audacity, it is a claim which is very far from a discredited prescientific myth. In fact, no observation has ever laid the presumption to rest. And today, four centuries after the scientific revolution, the doctrine is again reemerging. In these last decades of the twentieth century, its credibility is being enhanced by discoveries in several branches of fundamental science.

We have already seen the ID contention that if the universe weren't precisely as it is, human life would be impossible. As stunningly improbable as life may seem in physics and cosmology, it looks at least as unlikely at the chemical level where the naked eye can't go. And as audacious as the claims of Intelligent Design may appear to mainstream scientists, its proponents find implausible such manifest design could be "random" or merely "apparent."

Cooking Up Carbon for Life

Denton begins his chapter on the inherent design of our carbon-based universe with a famous quotation from physicist Freeman Dyson, published in *Scientific American* magazine: "Nature has been kinder to us than we had any right to expect. As we look out into the universe and identify the many accidents of physics and astronomy that have worked together for our benefit, it almost seems as if the universe in some sense must have known that we were coming."

The seeming chaos of exploding stars dangerously close to the earth is, on second look, a finely tuned "accident" that gives us what we need for life. Not only is carbon dependent upon the nuclear furnaces of stars, but stellar distances must be exactly right for solar systems like our own to form. And the energy with which carbon is produced affects the structure of particular atoms, which in turn affect the creation of carbon, oxygen, and other requisites.

Denton quotes physicist Fred Hoyle on the subject of carbon-based life. "A commonsense interpretation of the facts suggests that a super intellect has monkeyed with physics, as well as chemistry and biology, and that there are no blind forces worth speaking about in nature."

In the views of both Intelligent Design and mainstream science, the physical tolerances are tight. Indeed Denton notes if there were a wider range of constants that could sustain life, one could better argue that apparent "purpose" was no more than an accident. The fact that the physical requirements need to be so impossibly precise indicates the most successful theory is that of Intelligent Design.

Our Sublimely Engineered Water

In Chapter 9, we saw how utterly perfect water is for carbon-based life. For Intelligent Design as well as mainstream science, water is a prerequisite to life—even if, as mainstream science asserts, it is an accident of history. It is this "even if" with which ID takes dogmatic exception; for ID, water is so perfect, so sublimely engineered, it had to have been designed.

> **Evolutionary Revelations**
>
> Water's miraculous fitness for life is manifest in many directions. To name a few, water's thermal qualities sustain microscopic life, as well as play host to warm-blooded mammals, fish, and vegetative life. Our biosphere simply wouldn't exist without the role that water plays—stabilizing temperature and protecting the earth from otherwise fatal radiation.

Without water, human bodies (composed mostly of water) wouldn't exist. At the cellular level, water is the medium for organic life—without it there would be no living organisms with which to contemplate our existence. Water nourishes the food we eat and carries nutrients to our cells. Water circulates as blood and is the medium for cellular life and replication.

If there were no water, but only crystalline solids, the dynamic life of organisms would simply be impossible—trapped within a prison of structural rigidity. And if gas were the medium, its volatility could not sustain organic life. Thus, according to Denton, "If water did not exist, it would have to be invented."

"Nothing Better Than Air"

If carbon is the ideal chemical compound to enable organic life, and water is the ideal medium to interact with this carbon-based world, then perhaps some form of gas would be an ideal medium to disperse these atoms of carbon, giving physicist Robert Clark to argue there could be nothing better than air. Not only is air good at disseminating carbon, but oxygen interacts with carbon—producing water, carbon dioxide, and energy used in organic cells. Yet if there were much more oxygen than the 21 percent of air's composition, forest fires would be a constant threat, and little vegetation would survive.

Indeed, our atmosphere is an ingenious system: the oxygen, which we exhale as carbon dioxide, is converted back to oxygen by plants. If there weren't this interchange, and the atmosphere were strictly oxygen or carbon dioxide, the earth would either be consumed by fire, or animal life would suffocate.

If the viscosity and density of air were otherwise, human beings would cease to exist. Our need to breathe is perfectly met by the balance of the biosphere, miraculously nourishing the planet with our waste and cultivating human life. As Denton says, "By plotting all possible atmospheric pressures against all possible oxygen contents, it becomes clear that there is only one unique tiny area … where all the various conditions of life are satisfied …."

> **Evolutionary Revelations**
>
> The remarkable balance of the oxygen-carbon dioxide recycling machine appears to Michael Denton as "designed"—not only for the maintenance of the biosphere itself, but for the sustenance of living things.

Answers to Our Origins

If abiogenesis is possible, then the way is made clear for a strictly scientific explanation of life. Human beings can be seen as a natural product of material existence, with no need for intervention by any transcendent, "intelligent" agent. Answers to the question of whence we come naturally concern where we are going; so it is a question that most of us are inclined to take pretty personally.

It may be true that such a quest is finally religious. This is not to say that it has to be a metaphysical pursuit; it is rather to say that it has far more importance than, say, playing a game of cards. And indeed, that it is such a concern to neo-Darwinists indicates why abiogenesis is a critical part of their program. The chance origination of life from inorganic matter allows the unguided, material process that indeed is evolution.

Still, most ID scientists contend their critique of mainstream chemistry has little to do with metaphysics and everything to do with science. However mainstream chemists may "wave their hands," as Michael Behe describes, there is little to indicate that natural processes have achieved the complexity of chemical structures. For Intelligent Design, mainstream chemistry has yet to render a credible explanation; thus we shall see how, for ID proponents, chemistry manifests "design."

Pasteur Debunks Abiogenesis

Until his death, creationist Henry Morris lauded the work of Louis Pasteur as a definitive voice that put the question of abiogenesis to rest. Though not as critical an issue to Intelligent Design as it is to creationism, the question to ID proponents remains: Where's the beef? Where's the proof?

The great Louis Pasteur opposed the idea of spontaneous generation, leading him to test the hypothesis using several experiments.

Evolutionary Revelations

Nineteenth-century French microbiologist and chemist Louis Pasteur's work in bacteriology led to vaccines combating diphtheria, rabies, and anthrax. His name gave us the verb "to pasteurize," meaning to heat food to such a degree that dangerous organisms within the food are killed.

What is ironic is that Pasteur's opposition to the idea of spontaneous generation was not to defend God-fearing people against an enemy atheism. In fact, Pasteur opposed abiogenesis proponent Felix Pouchet who thought that abiogenesis happened by divine providence, attempting to seize the idea from materialists for whom God wasn't necessary.

Sterilizing Pouchet's flasks (in which Pouchet was claiming that spontaneous generation happened), Pasteur showed how they contained living organisms that had incubated microscopically.

Mainstream scientists have since said what Pasteur did not prove was that life could not come from inorganic matter. Pasteur may have shown the unlikelihood that such an event could happen, but this is far from saying that spontaneous generation was impossible. It is one thing to say it didn't happen in most cases, and quite another to say that it couldn't happen. Mainstream scientists claim that it needn't happen much to have produced human life as we know it.

Hurray for Miller-Urey?

The Miller-Urey experiments still stand at the center of the abiogenesis debate—50 years after graduate student Stanley Miller came up with his idea. As discussed, Miller sought to recreate the prebiotic atmosphere of the earth, thereby simulating the climatic conditions from which organic life might have come. At least, *might* have come—except for several issues that compromise the experiment.

First, oxygen was excluded from the atmosphere that Miller had created because it would have been lethal to the amino acids that he was trying to produce. Amino acids are the "building blocks" from which organic life comes, yet by scientific accounts, in all likelihood, there was oxygen in this atmosphere.

In *Evolution: A Theory in Crisis*, Denton says:

> In the presence of oxygen any organic compounds formed on the early Earth would be rapidly oxidized and degraded. For this reason many authorities have

advocated an oxygen-free atmosphere for hundreds of millions of years follow-ing the formation of the Earth's crust. Only such an atmosphere would protect the vital but delicate organic compounds and allow them to accumulate to form a prebiotic soup. Ominously, for the believers in the traditional organic soup sce-nario, there is no clear geochemical evidence to exclude the possibility that oxy-gen was present in the Earth's atmosphere soon after the formation of its crust.

If there were no oxygen, there would be another problem. Oxygen absorbs the ultra-violet light that emanates from the sun, thereby protecting organic matter from lethal radiation. If there were no oxygen, organic molecules would be unable to form—hog-tying scientists that either way, organic life is impossible.

Unlikely Beginnings

Aristotle believed that creation required a "pneuma," or vital force, which was already there in nonliving matter to make for living organisms. The appeal to the transcen-dent is made by Intelligent Design in a different way—in exploring the mystery of how organic life could come from inorganic chemicals. According to ID, the answer is, it can't, and that something else has to be invoked beyond random chance and pro-cesses accidentally aligned to enable this improbable life.

Given the improbability that life would come by purely material means, proponents of ID assert organic life cannot be materially explained. If ID fails to offer scientific explanations for how these structures came into existence, it makes minimalist claims of "Intelligent Design" manifest in chemistry. Despite the valiant efforts of main-stream scientists to garner naturalistic answers, the more we know of chemistry at the molecular level, the more it may appear to be designed.

God's Messenger

A frequent mainstream scientific assumption is that given enough time and know-how, science will be able to fill in the gaps in our knowledge of the physical world.

On this assumption, progress is science's friend, which turns away the ignorant past to let the light of knowledge reveal life's waiting truths that we once naïvely thought were "mysteries." Yet at the molecular level, say proponents of ID, life hasn't so revealed itself—except to express a complexity that none of us could have imagined.

Fact or Faith _____

Intelligent Design proponents say that as we have gradually come to understand the ways that living cells are put together, we can only be astounded by the "miraculous" engineering that is written into all of life. Consequently, so say ID proponents, we have little choice but to infer design. Additionally, the more we learn and know of cellular molecular complexity, the more we must entertain the prospect of a designer.

It is in such a context the ID argument is made from biological information.

We saw scientific problems in explaining how inorganic substances create amino acids, which are required to produce organic substances. Another problem is explaining how these substances, by which organic molecules are created, could be assembled into proteins, or chains of DNA, by purely natural means. In the face of the enormous information expressed in these finely engineered machines, it is difficult for advocates of ID to believe that it happened all on its own.

The Genetic Instructions for Life

As previously discussed, DNA (deoxyribonucleic acid) is an acid that contains the genetic instructions that determine biological development. Often referred to as "the blueprint of life," DNA is cellular material bearing the chemical information needed for organisms to grow and live.

The discovery of the DNA molecule (by Francis Crick and James Watson in the 1950s) has unlocked the mystery of human life at the molecular level. Yet as many questions as it answers, it asks—about the origin of the information borne in the human genome which allow human beings to develop in all their uniqueness. For Intelligent Design, DNA is proof of intelligence in the universe, whereby the startling complexity of this human life could only come by way of design.

The Origin of the Information?

Dr. Stephen C. Meyer, an ID advocate, points out that beyond their complexity, proteins are "specifically complex." Although the material stuff of proteins is simple amino acids, their function—what it is that they do—requires a specific sequence. If the sequence isn't just right, then the protein cannot perform its necessary function. Its construction must be as specific as the function that it is given to perform.

In *DNA and Other Designs* (First Things, 2000), Meyer writes:

> To form a protein, amino acids must link together to form a chain. Yet amino acids form functioning proteins only when they adopt very specific sequential arrangements, rather like properly sequenced letters in an English sentence. Thus, amino acids alone do not make proteins, any more than letters alone make words, sentences, or poetry. In both cases, the sequencing of the constituent parts determines the function (or lack of function) of the whole. Explaining the origin of the specific sequencing of proteins (and DNA) lies at the heart of the current crisis in materialistic evolutionary thinking.

The effort to explain protein sequences by general chemical laws seems, to most proponents of ID, to be inadequate. The nucleotide bases that determine the structure of the DNA molecule can bond with each other with equal likelihood to formulate the instructions.

The fact that they form particular instructions in the face of equal bonding begs the question: who or what put these particular instructions together? The reason, Meyer argues, that laws of chemical attraction are not adequate to the task is that they can't convey the enormous complexity that DNA requires. DNA structures are vastly more complex than is the molecular structure of salt (mentioned in Chapter 9). Such crystalline structures bear information which is highly repetitive, whereas the highly specified sequential information of DNA is unpredictable.

"Thus, mind or intelligence or what philosophers call 'agent causation' now stands as the only cause known to be capable of creating an information-rich system, including the coding regions of DNA, functional proteins, and the cell as a whole," continues Meyers. "Because mind or intelligent design is a necessary cause of an informative system, one can detect the past action of an intelligent cause from the presence of an information-intensive effect, even if the cause itself cannot be directly observed."

Meyers adds that because information requires a source of intelligence "the flowers spelling 'Welcome to Victoria' in the gardens of Victoria Harbor in Canada lead visitors to infer the activity of intelligent agents even if they did not see the flowers planted and arranged."

The Least You Need to Know

♦ One of the unwritten precepts underlying many ID arguments is that the finely tuned universe is fine-tuned for the sake of human beings.

♦ Carbon, water, and air appear to be sublimely engineered.

♦ For many ID advocates, scientist Louis Pasteur debunked "abiogenesis," which means the spontaneous generation of life.

♦ The specifically complex construction of DNA is, for many proponents of ID, evidence of intelligence in the chemical constitution of life.

13

Intelligent Design Biology

In This Chapter

- ◆ Irreducibly complex
- ◆ Objectively seeing "design"
- ◆ The geneticist's problem
- ◆ Testing the past

Biology—the name stemming from the ancient Greek word *bios*, which in English literally means "life," and from *logos*, also from the Greek, meaning "word," or perhaps an object of "study"—is the scientific discipline that embraces living organisms as, in short, "the study of life." This linguistic reading makes academic sense. Yet in the vocabulary of Intelligent Design, "biology" could also be seen in the metaphorical sense of "word of life"—with its theological implications.

If you remember, Aquinas took the apparent design that he saw in the universe as a sign of the presence of an engaged designer, and, ultimately, of God. One could argue that for many theistic proponents of Intelligent Design, the miracle of biological life is the living word of God. Indeed, as we have seen in ID chemistry, and specifically the work of Michael Denton, this is an "anthropocentric" (human beings at the center) universe.

Over and against mainstream science's assumption that human beings are nothing special, most proponents of Intelligent Design believe design is for a human purpose. They believe that many biological structures manifest the hand of a designer, whose plan has always been to create and sustain the miracle of human life. And as we shall see, Intelligent Design covers the biological landscape: from cellular machinery at the molecular level to species spanning millions of years.

We will look at three examples of what ID proponents believe express Intelligent Design. The construction of the eye, the blood-clotting mechanism, and the bacterial flagellum are testaments, according to Michael Behe, of material design. From there we will look at questions ID poses about the theory of evolution, which most ID biologists reject as an "inadequate explanation" of life. And finally, "gaps" in the fossil record are brought to light by ID advocates—all contributing to a perspective that questions the mainstream scientific paradigm.

What in the World is Irreducible Complexity?

We have heard the term irreducible complexity before (certainly in this book). The term itself sounds "complex," and perhaps too lofty or pedantic. But it's actually a very simple concept.

Blinded By Science _____

By irreducible complexity I mean a single system composed of several well-matched, interacting parts that contribute to the basic function, wherein the removal of any of the parts causes the system to effectively cease functioning. An irreducibly complex system cannot be produced directly (that is, by continuously improving the initial function, which continues to work by the same mechanism) by slight, successive modifications of a precursor system, because any precursor to an irreducibly complex system that is missing a part is by definition nonfunctional.

—Michael Behe, *Darwin's Black Box*

This may be one of the most frequently quoted passages of Michael Behe's best-selling book *Darwin's Black Box*, if not one of the most contentious concepts of the ID movement.

Behe introduced the term irreducible complexity in 1996. Since then it has become an icon in the ID movement—invoked by proponents and disputed by opponents in this scientific debate.

One of the reasons it is so hotly contested by mainstream biology is that it takes direct aim at the Darwinian theory of evolution. Indeed Behe asserts if it could be shown that there exists such a thing as an irreducibly complex biological system, it would be "a powerful challenge to Darwinian evolution." For in *The Origin of Species* published in 1859, Darwin himself said, "If it could be demonstrated that any complex organ could not possibly have been formed by numerous, successive, slight modifications, my theory would absolutely break down."

Behe believes this has been demonstrated. In the cases of the human eye, the blood-clotting system, and the bacterial flagellum, evolutionary explanations of these organisms make little sense to him. Before we look at each particular case, we shall review Behe's infamous example of the mousetrap—which he uses to explain irreducible complexity in biology.

Behe explains that the mousetrap is composed of several mechanical parts: the wooden base, the catch, the spring, and the hammer. He goes on to say that all these parts must be functional in order for the mouse trap to work. Take one part away, and the mousetrap breaks down. Take away the base, and the spring has no anchor by which the hammer can squash the mouse. Take away the spring, and the hammer rests inert on a useless wooden base. Take away the catch, and there is no way for the hammer to be sprung, and so forth. All the pieces must be engineered as a single system for the mousetrap to operate.

> ### Evolutionary Revelations
>
> Evolution says organisms evolve not by prior design, but by the need to be optimally fit in order to better serve the organism. Because none of the parts can be functional until the whole system operates, the evolution of these parts makes no sense in the absence of a prior purpose.

Biological systems are also comprised of interdependent parts, and all must operate together in order for the whole system to function.

In Darwin's words, if there is no need for "numerous, successive, slight modifications," then there is no evolution. So for Behe, the question is, how can the mousetrap know it will be a mousetrap until it is all together; how can an organism know what it will be until it has already happened? How can any "whole"—mousetrap or organism—evolve toward a functional purpose until it knows what that purpose is: trapping mice, seeing, clotting, swimming?

ID's Poster Child

We have heard from Kenneth Miller that the bacterial flagellum is ID's "poster child." Until the advent of the electron microscope, little could be known of the cell.

According to Behe, the cell was a "black box," and its contents were unknown. But since the 1950s, technology has opened up the black box—allowing researchers to plunge the once-unknown workings of the cell.

For Behe, among these once unknown contents, is the bacterial flagellum. As we've discussed, this flagellum is a swimming mechanism that propels bacteria through water. Embedded in the membrane of the cell, it has more than 40 interactive parts—from a filament propeller, to stators, rods, and hooks, and a host of chemical substances.

As a paddling mechanism (remember, "the little engine that could"?), the flagellum requires three major moving parts—a paddle, a motor, and a rotor—in consort with complex submechanisms. By virtue of the interdependent nature of these mechanisms, Behe finds it difficult to imagine how the flagellum could have possibly evolved. If evolution takes place by natural selection of the fittest interactive parts—but the parts are useless until they are part of the whole aquatic machine—then either it was one fortuitous mechanical accident or the parts somehow "knew" they had to come together and were part of a grander design.

Thus, Behe says, the bacterial flagellum is irreducibly complex. Evolution argues the only purpose in organic change is physical survival, and yet the flagellum cannot survive until it is all of a piece. So it would seem that there was some designer who had already conceived of a purpose, which, after it was put together, enabled bacteria to swim.

Blinded By Science

… as biochemists have begun to examine apparently simple structures like cilia and flagella, they have discovered staggering complexity, with dozens or even hundreds of precisely tailored parts … As the number of required parts increases, the difficulty of putting the system together skyrockets, and the likelihood of indirect scenarios plummets … Darwinian Theory has given no explanation for the cilium or the flagellum. The overwhelming complexity of the swimming systems push us to think it may never give an explanation.

—Michael Behe, *Darwin's Black Box*

The Clotting Cascade

In Chapter 10, we saw the basics of the human blood-clotting system as described by evolutionary biologist, Kenneth Miller. One might argue that Miller had a motive for this simplified treatment of the mechanism. If it could be kept simple, it might be clearer to others how it could have evolved, rather than portraying it as "irreducibly complex."

So the approach of Intelligent Design stresses complexity in its biochemical under-standing of the blood-clotting cascade. To this end, Behe likens the clotting cascade to a Rube Goldberg cartoon, in which absurdly elaborate, nonsensical machines operate with comical perfection. The nonsense aside (given the life and death stakes of the blood-clotting process), Behe describes the clotting cascade as irreducibly complex.

What does this mean? It is complex insofar as it is a system that uses many parts to perform a function, and irreducible in that taking one part away makes it cease to function. But in this case of blood clotting, it is sequential—one thing makes another thing work—and so requires a dynamic precision that, to Behe, implies design.

For ID, the mainstream biological explanation once again does not make sense. Although it might be true that the blood clotting process could occur another way, any evolutionary change, Behe argues, would incapacitate the system. As such changes occurred, the existing system would become nonfunctional, and therefore this Darwinian scenario would undermine its own "survival."

Returning to the Rube Goldberg cartoon, ID contends that the blood-clotting system is staggeringly complex. The process of clotting—just at the right place, and for just the right amount of time—involves scores of biochemical courses, all critical to its success.

Behe explains:

> When an animal is cut, a protein called Hageman factor sticks to the surface of cells near the wound. Bound Hageman factor is then cleaved by a protein called HMK to yield activated Hageman factor. Immediately the activated Hageman factor converts another protein, called prekallikrein to its active form, kallikrein. Kallikrein helps HMK speed up the conversion of more Hageman factor to its active form. Activated PTA in turn, together with the activated form of another protein ... called convertin, switch a protein called Christmas factor to its active form. Finally, activated Christmas factor (which is itself activated by thrombin in a manner similar to that of proaccelerin) changes Stuart factor to its active form.

You get the point. So does mainstream biology: the conflict lies in its explanation. But Behe serves as the voice of ID when he says, "Slogging through a description of the blood clotting system makes a fellow yearn for the simplicity of a cartoon Rube Goldberg machine."

The Eyes Have It

Perhaps the most tangible example of irreducible complexity is found in the human eye. Mainstream biology answered the question, "What good is half an eye?" Now it is ID's turn.

Darwin was acutely aware of this problem. In fact, Darwin addressed the quandary in a section of *The Origin of Species*. After all, if eyes appeared out of nothing in one-stop miraculous fashion, then gradual evolutionary development would be unnecessary.

Given the state of biological research in the 1850s, Darwin was unable to explain how the eye might develop—let alone, evolve. So he turned to examples in various species that exhibited simpler eyes as a way to credibly hypothesize gradual development.

Fact or Faith

Looking at a variety of eyes—and using examples of eyes from what he said were "light-sensitive spots" to "intermediate eyes"—Darwin was able to show the complex human eye as a product of evolution.

For Behe and most ID advocates, Darwin's eye answer is unsatisfying because it doesn't explain how the eye can see in the first place. However, Behe excuses Darwin's inability to explain the visual specifics, given that he was too early to grasp it at the molecular level.

As a biochemist, Behe understands the eye at a molecular level. He understands the literally hundreds of sequential chemical processes required to create an image that is comprehendible to the human mind.

Vision at a cellular level was a "black box"—an unknown mechanism—that has since been opened by the dramatic advances of biochemistry.

"Now that the black box of vision has been opened, it is no longer enough for an evolutionary explanation of that power to consider only the anatomical structures of whole eyes, as Darwin did in the nineteenth century (and as popularizers of evolution continue to do today)," writes Behe. "Each of the anatomical steps and structures that Darwin thought were so simple actually involves staggeringly complicated biochemical processes that cannot be papered over with rhetoric."

Specified Complexity

If ID proponents are claiming the presence of design in the universe—and specifically, design in organic systems not accounted for by random processes—how do we determine what has been "designed," and what has been randomly produced? In biological terms, how do we know an organism was produced by a designer?

Enter Bill Dembski's "specified complexity." Specified complexity is a concept that attempts to discern the presence of design. Rather than simply determining design by subjective intuition, Dembski claims that it is possible to objectively ascertain design.

You'll remember Michael Behe's biochemical concept of "irreducible complexity," which, when determined to be present, he believes indicates design. Dembski understands "IC" as a subset of "specified complexity," wherein the interacting parts of the bacterial flagellum are complex in a specific way. It is important to note that this concept takes aim at a basic evolutionary tenet: that random processes, in the form of mutations, produce only "apparent" design.

Specific and Complex

In Dembski's mind, it's not enough that an organism is complex for it to be designed. Random word generation, that creates a random sentence, is inherently complex—but in all likelihood, it comes out as meaningless babble, without specific meaning. On the other hand, a Shakespearean sonnet is complex, and specifically so: rendering images, evoking sensibilities and meanings that were clearly designed.

Dembski cites the movie *Contact* (based on Carl Sagan's novel) to show how specified complexity has been used in scientific explorations funded by the government. The movie was based upon SETI, "Search for Extraterrestrial Intelligence," the scientific program where technology was used to search the galaxy for signs of intelligence. The assumption was that complex radio signal patterns, when judged to be specifically determined—in this case, as a sequence of beats and pauses corresponding to a series of prime numbers—would be a sign of alien intelligence that was present in the universe (or for the sake of Dembski's argument, of the presence of Intelligent Design).

Blinded By Science

SETI or the "Search for Extraterrestrial Intelligence" is a scientific program funded in part by the U.S. government. Among SETI's projects is a multi-directional monitoring of the skies for deliberate or "intelligent" communications transmissions from unknown worlds.

The SETI program was based on the fundamental assumption that specified complexity is not only a sign of intelligence, but that its likelihood is measurable. In this regard, not only is the bacterial flagellum irreducibly complex; Bill Dembski would claim that its specified complexity couldn't have been made by random process. However well we know in our daily lives that something must have been designed, within the realm of science, somehow it is "illegal" to make such a determination.

Specified Complexity In Every Day Use

Dembski describes a human being's production of something that has been designed. He first decides upon a purpose that has to be accomplished, and then forms a plan to accomplish it. Gathering materials, he uses the plan to finally create the designed object. "In the case of human designers, this ... process is uncontroversial. Baking a cake, driving a car, embezzling funds, and building a supercomputer each presuppose it."

Dembski says we can work back from an object we have found to determine if it was designed. So if we found a computer lying on the ground (perhaps like William Paley's watch), we would probably determine that it was in fact, designed, rather than randomly produced. Though we entertain the presence of design on the mundane landscape of our everyday lives, Dembski says within the scientific arena, this is deemed inappropriate.

"The exclusion of design from biology certainly contrasts with ordinary life where we require three primary modes of explanation: necessity, chance, and design. Nevertheless, in the natural sciences one of these modes of explanation is superfluous, namely, design. From the perspective of the natural sciences, design, as the action of an intelligent agent, is not a fundamental creative force in nature. Rather, blind natural causes, characterized by chance and necessity and ruled by unbroken laws, are thought sufficient to do all nature's creating." (William Dembski, in his article, "Detecting Design in the Natural Sciences.")

Thus Dembski pits himself against Darwin's contention that human beings can be explained as the product of an unbroken natural process—without the need of a designer. He objects to the idea that invoking a designer takes us from the realm of science. Insofar as SETI recognized the chance of an intelligence, and therefore, of some kind of design, he believes there's little difference between these enterprises as scientific exploration.

The Need for Objectivity

To the argument that ascertaining design is inevitably subjective, as a mathematician Dembski responds that there are ways to ensure objectivity. Employing a "probability bound"—which seeks to describe the possibilities for all creation in the universe—he believes that his method of determining design can escape unscientific conclusions. Dembski has determined a number that describes the degree of improbability that something was created by a blind, random process, rather than by being designed.

In Dembski's mind, it is critical that such a measurement be impartial—that it not look back to rationalize an answer, but look ahead to what the stuff of life reveals. "The important thing about specifications is that they be objectively given and not just imposed on events after the fact. For instance, if an archer fires arrows into a wall and then paints bull's-eyes around the arrows, the archer imposes a pattern after the fact. On the other hand, if the targets are set up in advance (specified), and then the archer hits them accurately, we know it was by design." ("Detecting Design in the Natural Sciences")

While there are many criticisms of specified complexity as formulated by Dembski, it would seem that like other criticisms of ID, most have to do with the fear of "a creator." Insofar as objectivity is the goal of science, mainstream scientists foresee a danger in considering the presence of created design, and therefore a designer. However, Dembski says identifying a designer isn't a part of the program, and that the ID point of view must stay focused on "the stuff"—which he believes to be embedded with design.

Again, in his article, "Detecting Design in the Natural Sciences" (*Natural History*, April 2002) Dembski writes:

> Design has had a turbulent intellectual history. The chief difficulty with design to date has consisted in discovering a conceptually powerful formulation of it that will fruitfully advance science. It is the empirical detectability of intelligent causes that promises to make the theory of intelligent design a full-fledged scientific theory and distinguishes it from the design arguments of philosophers and theologians, or what has traditionally been called 'natural theology.'

Haldane's Dilemma

Haldane's dilemma is, as one might guess, a dilemma posed by (J. B. S.) Haldane. In 1957, this famous geneticist wrote a paper, "The Cost of Natural Selection." The dilemma he describes essentially calls into question the possibility that evolution could have evolved human beings from a common ancestor.

It should be noted that not every ID proponent sees this dilemma as a problem. Nonetheless, for many, Haldane's dilemma remains a dilemma unresolved and figures into ID skepticism about evolutionary mechanisms.

It should also be noted that this is a criticism shared with some creationists. What distinguishes the two positions is their principal motive: for creationists, it is to reject common descent in favor of biblical creation. For Intelligent Design, the motive is more complicated, given its ambiguous views on evolution as a theory to be judged on scientific grounds.

Not Enough Time

Haldane's dilemma assumes that an organism gradually evolves in response to changes in the environment where the organism lives.

> ### Evolutionary Revelations
>
> Michael Behe has stated that as far as he can see, common ancestry makes sense—like all large movements, Intelligent Design bears a variety of viewpoints.

As a geneticist, Haldane obviously knew about "genes" which make up all organisms—but at the time of Darwin, genetic research was a century away. So the dilemma is based on evolution at the genetic level, and it takes place when a gene is substituted to maximize the organism's survival.

What troubled Haldane was that, given the assumptions of evolutionary theory, there wouldn't be enough time for humans to evolve from a so-called "common ancestor."

For a mammal to evolve, it must receive beneficial genetic substitutions, and Haldane showed the rate of beneficial substitution to be 1 in 300 generations. If it took 10 million years for the common ancestor to evolve into a human (as is commonly accepted, though many estimates are significantly less than this), then according to Walter ReMine, author of *The Biotic Message*, assuming 20-year generations, 1,667 substitutions would be possible.

What is the significance of this calculation? Steeped in evolution, Haldane was concerned that there simply wasn't enough time for an apelike ancestor to have genetically evolved into an upright, developed human being. Unlike creationists, who appeal to Genesis to have answered how humans were created, Haldane invoked scientific assumptions to question common descent.

This is not to say that those who are convinced that there is indeed a dilemma reject the theory of evolution as a scientific explanation. Indeed, Haldane himself was no creationist but a thorough-going scientist who was asking hard questions within the scientific context of evolutionary assumptions.

Based on these same evolutionary assumptions—granting microevolution, yet questioning its capacity to form human beings—Walter ReMine said:

> … evolutionists must accept what nature doles out—and we can observe what nature doles out …. Take the blend of beneficial substitutions observed in nature. The coloring of melanic moths and the beak size of Galapagos finches are noted examples. Those represent the basic building blocks that nature has to work with, the creative "power" of nature. Leading evolutionists acknowledge (indeed, most of them insist) that substitutions with small effect predominate in evolution. Are 1,667 substitutions (like those) sufficient to create all the human adaptations? The tripling of brain size, fully upright posture, language, speech, hand dexterity, hair distribution, and appreciation of music, to name a few.

Evolutionary Revelations
Briton J. B. S. Haldane was an evolutionary biologist and geneticist and one of the founders of the study of "population genetics." A professed communist, he served in the famous Black Watch during World War I, and wrote, "If I live to see an England in which socialism has made the occupation of a grocer as honourable as that of a soldier, I shall die happy."

More Questions Than Answers

We saw how mainstream biology holds to a number of assumptions: macro- and microevolution, speciation, and common descent. Although it is impossible to generalize a single ID position with regard to these ideas, ID resists evolution as an overarching explanation for life.

Haldane's dilemma neatly packages this resistance in scientific terms—allowing microevolution but calling into question the dynamics of macroevolution.

ID's view of speciation is ambiguous—although doubtful to some, it's not a problem for others—whereas still for others, it's irrelevant.

The same goes for the issue of common descent, despite mainstream scientific claims that Intelligent Design is in fact "creationism" behind a pseudoscientific mask. The truth is that ID, regardless of how one comes out on the issue of evolution, is principally asking some reasonable scientific questions that deserve scientific answers.

Fortunately for science, the slide rule and the laboratory are not the only paths to knowledge. The earth is the classroom for paleontologists: who dig in the mud, work in the field, and burrow through a buried past. Fossils, found in the four corners of our world, provide critical evidence: the only problem is that for the data they provide, it is human beings who must interpret them.

Fossils for Testing

For many proponents of Intelligent Design—unlike strict creationists—the fossil record is secondarily important or even irrelevant. Yet insofar as science is empirical, we must dwell within the realm of the concrete; indeed, the fossil record provides a critical benchmark by which to test evolution. Specifically, Darwinian evolution as a biological explanation depends upon the presence of certain fossils to vindicate its scientific claims.

As we shall see, Darwinian evolution presumes gradual change of organisms—change that is "selected for" to maximize survival in their various environments. It is therefore logical to expect that in the dirt of an evolutionary fossil record, one would find evidence of gradual, continuous change from a common ancestor. Because speciation is presumed to occur, the presence of transitional fossils from one species to another is, again, a logical expectation.

Moreover, these biological changes must occur within evolution's constraints: populations, reproduction and substitution rates, and other factors seen in Haldane's dilemma. ID proponents generally acknowledge the quirks of fossilization and generally don't insist on a "perfect record" to indicate evolution. Still, Intelligent Design desires a greater explanation than evolution alone—unlike mainstream biology, for whom evolution alone is enough.

Criticizing the Gaps

It is worth mentioning that creationist critiques of "gaps" in the fossil record are used to vindicate biblical creation as an explanation for the source of life, whereas ID proponents express their skepticism in terms of inadequate scientific evidence aimed at vindicating evolution. So they engage the question of the time required to evolve organisms as a possible contradiction to evolution's own presuppositions.

Instead of treating the record's incompleteness simply as a negative critique—that is, the absence of transitional fossils proving that evolution is "wrong"—Intelligent Design would be more inclined toward a predictive hypothesis that would catalyze discussion about the scientific problems rather than the gaps in our knowledge.

Lack of Transitional Fossils

As previously mentioned, whereas creationists are more deeply invested in the absence of transitional fossils, for many proponents of ID this is of secondary concern. Yet insofar as ID dovetails neatly with creationists' skepticism about evolution, ID advocates are often inclined to embrace the evidentiary support that the fossil record provides. For however perfect any fossil, or however perfectly it fits into a historical context, it will always be a matter of interpretation—which is where the fun begins.

Modern bats are thought to be almost identical to their ancient ancestors and, to some, do not appear to have transitioned from four-footed predecessors. According to the Berkeley Museum of Paleontology, "[the earliest complete bat] fossils represent essentially modern-looking microchiropterans; bats had evolved all of their characteristic features … in fact, the oldest known complete fossil bat … shows specializations of the auditory region of the skull that suggest that this bat could *echolocate*. Thus, as convinced as Darwinist Richard Dawkins is about the evolution of bats, there are those who believe there are no transitional forms leading to the 'modern bat.'"

Fact or Faith _____

An ID hypothesis might look like this: species that are designed will not show signs of evolution in the record. Thus, if transitional fossils are discovered somewhere within the fossil record, then evolution would be vindicated, and the ID hypothesis that such species were designed, disproved.

def•i•ni•tion _____

Echolocate (or echolocation) is the method of locating objects and points in the distance by determining the time for an echo to return and the direction from which the echo came. Bats echolocate as a means of navigating from one point to another.

In the same way, whales, often touted as an icon of transitional fossils, have been questioned for their alleged "hind legs," which are supposed to show that they were once amphibious. Some critics cite the absence of a pelvis (which would have been required for them to walk) as a sign that whales are not an example of evolutionary speciation.

It has been suggested that these vestigial "legs" were used for copulation, or may have aided them in swimming and living in an exclusively aquatic habitat.

The same arguments apply to other transitions: the appearance of fish, of fish to amphibians, of amphibians to reptiles, and reptiles to mammals. All hotly debated, and paleontologist Stephen Jay Gould implies that given the speculative nature of the record, such debate may never cease.

"The absence of fossil evidence for intermediary stages between major transitions in organic design, indeed our inability, even in our imagination, to construct functional intermediates in many cases, has been a persistent and nagging problem for gradualistic accounts of evolution," writes Gould in "Is a New and General Theory of Evolution Emerging?" (*Paleobiology*, January 1980).

500 Million Years Ago

The first 30 years of the Cambrian period (570 to 500 million years ago) saw a veritable explosion of plants and animals as has never been seen again. The sudden appearance of myriad forms of life with no apparent ancestor has captured the imaginations of those who believe that organisms are designed. Worms, mollusks, sponges, even vertebrates, seem to have appeared out of nowhere. As neo-Darwinist Richard Dawkins said, "It's as though they were just planted there, without any evolutionary history."

In some ways, the Cambrian explosion is "tailor made" for proponents of Intelligent Design. With no apparent predecessors, and in the blink of an eye (10 million years is a blink), organisms abruptly appeared as if to undermine the thought that they had been evolved. One could imagine how this eerie history of the sudden appearance of life also perfectly fits with the Genesis story found in the Bible.

Punctuated equilibrium notwithstanding, the "evolutionary plot" may seem to tighten—and for those who are adherents to Intelligent Design, it tightens around the throat of evolution. Or does it? As much as we have heard about "Darwinian evolution," we rarely hear directly from Darwin—most of us assume that "neo-Darwinism" is just Darwinism for the modern day. This may not be the case, as far as we have come from Darwin's *The Origin of Species*; it would be refreshing, and perhaps even surprising, to review what Darwin himself actually said.

The Least You Need to Know

◆ Most theistic proponents of Intelligent Design see biological life as a living word of God. Most nontheistic proponents of ID still see design as manifest in biological structures.

◆ Irreducible complexity points to a single system composed of several well-matched, interacting parts that contribute to the basic function, wherein the removal of any of the parts causes the system to effectively cease functioning.

◆ ID's poster child is the bacterial flagellum (the little engine that could).

◆ Haldane's dilemma calls into question the possibility that evolution could have evolved human beings from a common ancestor.

◆ The Cambrian explosion—the sudden appearance of life on Earth between 570 and 500 million years ago—for many fits with the Genesis account of the birth of life.

Part

Darwin's Meaning of Life

Charles Darwin's theory of evolution is one of ID's primary hurdles.

Beginning with a look at just who this nineteenth-century naturalist was, and how he changed the generally design-inclusive approach to science, we move into the theory's ideas of mutation and natural selection, and consider the far-reaching implications of Darwin's theory today.

14

The Famous and Infamous Charles Darwin

In This Chapter

- ◆ The importance of Darwin
- ◆ Darwin's science
- ◆ The early impact of Darwin's science
- ◆ Darwin's science today

There is something less personal about the starry heavens than our human origins. As intriguing as the cosmological constants that miraculously allow carbon-based life may be, they are somehow removed from the stuff of the flesh that sustains our consciousness. It is when we explore the mysteries of where we come from and where we are going that we wake up and take notice of the metaphysical questions that evolution implies.

For some of us, the most important question is the existence of God. For the more agnostic thinkers of our number, the critical question may be just how nature works. And still for others of us—of which the Intelligent Design proponents might be counted—the question is how material life points to a designer.

So how is Darwin at the heart of the Intelligent Design debate? In the next chapter, we focus on the content of Darwin's theory; but for now, we look at Darwin the man and his place in our history. We also consider the culture in which his thinking was conceived, and the impact it had on his nineteenth-century British surroundings.

Why Darwin?

Few scientific luminaries have had the same level of impact on the world as that of Charles Darwin. In fact, during a recent talk-radio show, James Watson, of DNA fame, and E. O. Wilson, the eminent evolutionary biologist, were both asked who they believed was the most important person in history. Without missing a beat, they both agreed that it had to be Charles Darwin.

Impact aside, it could certainly be argued in many scientific and sociological circles that few—if any scientific luminaries—have stirred as much controversy as Darwin. Indeed, one could argue that Darwin and his so-called scandalous ideas undergirding evolutionary theory are at the bottom of the current debate about Intelligent Design. The meat of the matter for most of us is the purpose of our lives; and no matter how we might perceive that purpose, Darwin has shaped the way we think about it.

Darwin the Man

Darwin was born in 1809 in Shrewsbury, Shropshire, England. One of six children, he was the son of a prosperous physician, Robert Darwin, and Susanna Wedgwood Darwin, a member of the Wedgwood family who had made china famous.

When young Darwin was 16, he apprenticed with his father caring for the poor in Shropshire, and went on to study medicine at the University of Edinburgh.

Fact or Faith

It was not unusual for clergy to be steeped in the natural sciences, seeking signs of the divine in the biological world around them. Until this time, biology was an amateur's discipline, and the church was an appropriate context for biological study.

Rather than being an avid student, as one might imagine, Darwin appears to have been disinterested in things that were considered to be conventionally academic.

In her book *Of Moths and Men*, Judith Hooper notes, "From an early age he was bored by school, and all his passions were directed to the outdoors, to fishing, collecting, hunting, and reading nature books. 'You care for nothing but shooting and dogs and rat-catching,' his worried father upbraided him, and it was true that he seemed destined for nothing more than country-squire obscurity. In his older brother's footsteps, he had been sent to the University of Edinburgh to study medicine, but he proved too squeamish, if not inept, and in 1828 was packed off to Cambridge to study theology instead."

What might seem ironic—that Darwin would, one day, be presumed anti-religious—was in fact a logical academic choice for most any lover of nature.

The theological tide was soon to turn when Darwin boarded the HMS *Beagle*, changing not only Darwin himself, but the very nature of science.

Darwin's Great Adventure

Darwin's voyage to the Galapagos Islands is the stuff of which myths are made. Often cited as the place where he glimpsed the truth of evolution, it was only several years after he returned that he came to this conclusion. If a theological education was in fact his father's choice, Darwin was as steeped as anyone else in his theistic Victorian culture.

Thus every organism was a gift of God, whose nature was ordained by its creator; and just as for many Christians now, any other perspective was a threat to the certainty of God. From rhinoceros to finch, biological life expressed God's immutable order, and their scientific study was undertaken in the spirit of revealing God's divine expression.

But something happened when Darwin agreed to accompany Captain Fitzroy on a voyage to South America, and to the Galapagos Islands. There, Darwin traveled far and wide collecting fossil specimens. Some of his geological writings were sent home to his botanist mentor, who publicized his findings and work to other British naturalists. By the time he returned, Darwin already enjoyed an academic reputation, inspiring his father to set him up as a "gentleman scientist."

Darwin's Evolution

An implication of believing that every organism was the product of God's handiwork is that species were "fixed"—not subject to change that implied they had a life of their own.

The trouble for Darwin—and it would mean trouble—was the evidence said otherwise: his specimens revealed the existence of species that were distinct, yet clearly related. He therefore conjectured that new species could develop by cumulative adaptations occurring in response to environments, which for whatever reason had changed.

Blinded By Science _____

In the light of these clear relationships, Darwin came to believe that all species must have originally descended from a common ancestor. What he had yet to figure out is what caused the biological changes that resulted in the emergence of distinctly novel species. The answer seems obvious to those of us who are steeped in evolution's assumptions: variations in organisms are inherited that maximize their chance of survival.

Darwin was affected by an essay on human populations by Thomas Malthus. Malthus noted that when populations go "unchecked," they increase exponentially; thus organisms are born into a necessary fight for their own physical survival. In Judith Hooper's words, "If unhindered by war, famine, or disease, the population would quickly outgrow its means of subsistence. Because more individuals are produced than can possibly survive, Darwin reasoned that among all organisms ceaseless competition is the name of the game, and small differences assume life-and-death importance. Birds with a certain shape of beak, the 'swiftest of slimmest grey wolves', or bees with some minute difference 'in curvature of length of proboscis, far too slight to be appreciated by us' will win the race for existence."

"Descent with Modification"

Up to this point, Darwin had no natural examples of this "natural selection." But artificial selection was being employed in horse and cattle breeding. From there, by the engine of the fight to survive, Darwin speculated that by natural means, traits were selected that maximized survival.

Yet more than simply selecting for strength, or admirable aesthetic characteristics, Darwin saw nature scanning every hidden feature of nature. Any variation that enabled an organism to improve its potential to survive would be inherited, thereby explaining biological "descent with modification." This idea quickly moved beyond a description of the evolution of species to one that explained the inherent nature of the whole of organic life.

Origin of Species

Darwin's theory was developed in 1839, but he did not publish the theory until 20 years later. When he received a letter from A. R. Wallace about Wallace's own evolutionary theory, Darwin realized that "survival of the fittest" applied to the academy, too. It was agreed that these two men would present papers several months later on their similar theories; because of the death of Darwin's infant son, however, Darwin was unable to.

The next year, Darwin published *On the Origin of Species by Means of Natural Selection.* Though a year earlier, the papers the two men had written stirred little controversy— perhaps because there had been other forms of evolutionary theory around—when *Origin of Species* was released in November 1859, the printing of 1,250 copies was already oversubscribed.

Darwin's Immediate Impact

Darwin's fear of the conflict (and he believed his theory was likely to produce conflict) came home to roost in his Victorian culture. It was just as he had predicted. If the topic of "evolution" could give rise to such passionate resentments in our time, imagine how it might have felt to the father of the theory in Victorian England.

Darwin was caricatured as an ape in cartoon depictions in magazines and was broadly ridiculed for the distressing idea that human beings were not formed by God. Darwin saved thousands of clippings, evidence that the controversy did affect him; and given his wife's own religious faith, it was no doubt a problem at home. Still he had many advocates of evolutionary theory; and as is the case in every controversy, an opposition mounted.

> **Evolutionary Revelations**
>
> Darwin was plagued by illness throughout his scientific career, and there is much speculation that it was due to the stress that the conflict engendered.

If this historic controversy seems oddly familiar, perhaps it is no accident. For some, the current Intelligent Design debate bears creationist overtones. One might also argue that Darwin was distorted by his followers in the way ID is presently distorted by creationist supporters.

"Darwin's Bulldog"

Renowned anatomist Thomas Huxley was instantly Darwin's advocate. In fact, when he first heard the idea, he purportedly replied, "How stupid of me not to think of that!" His wit and scientific expertise made him an ideal spokesman; so by his willingness to tangle with the church, he was known as "Darwin's bulldog."

The most celebrated exchange with the opposition was with Bishop Wilberforce of Oxford. At Oxford's University Museum, the stage was set for what would be a celebrated showdown between this religious creationist and Huxley, a self-proclaimed atheist.

The story goes that Wilberforce asked Huxley whether his monkey ancestry was through his grandfather or his grandmother, to which Huxley is said to have muttered that "the Lord" had delivered (Wilberforce) into his hands. Then, according to Hooper, Huxley replied, "If the question put to me is 'Would I like to have a miserable ape for a grandfather, or a man highly endowed by nature and possessed of great means and influence, and yet who employs these faculties and that influence for the mere purpose of introducing ridicule into a grave scientific discussion'—I unhesitatingly affirm my preference for the ape.'"

In any case, by the 1870s, most educated people accepted Darwin. His theory successfully explained a number of scientific facts: specifically, relationships between the species and the progressive nature of the fossil record. But as we shall see, for reasons that lay beyond the science itself, "descent with modification" would also become a thriving social theory.

The Fire of Social Darwinism

The idea of evolution fit well with a progressive view of social history, whereby the stronger prevail against the weak in the wider social arena. Although "survival of the fittest" is most often linked with the name of Charles Darwin, it was a contemporary philosopher, Herbert Spencer, who actually coined the phrase. Taking Darwin's idea of evolution by environmental adaptation, Spencer translated Darwin's theory into Victorian culture at large.

As industrialization brought an unprecedented sense of possibility and progress, Social Darwinism made common sense in Britain and America. The "invisible hand" philosophy of Adam Smith—whereby it was believed that economic selfishness rendered benefit to all—was a perfect counterpart to Darwin's implication that life was a struggle for existence, and that by such struggle individuals realize their optimal condition. Although this deviated from Darwin's belief that

> **Evolutionary Revelations**
>
> There was a dark side to Social Darwinism. The weak were to be "weeded out" by virtue of this law of nature, and therefore personal and economic strength were given moral justification. In its extreme, Social Darwinism was used to abuse the "lesser races," feeding into such philosophies as those of Nazi Germany.

evolution wasn't progressive, it nourished the culture's desire to realize the fruits of capitalism.

In America, the philosophy also made sense. It would be used to justify low worker compensation and inhumane working conditions, as well as to withhold philanthropic giving to those who were less fortunate. Industrialist Andrew Carnegie created hundreds of libraries to improve the lot of those who were already inclined to benefit from higher learning. Indeed it lent credence to American colonial exploits around the globe, invoking the ethic of "survival of the fittest" by way of military action.

Darwin's Decline

It may be hard to believe that Darwinian theory was ever in decline. The evolutionary view of biological life seems so unquestioned now that one might easily imagine from the start it was nothing but a winner. Even present anti-Darwin sentiment is less a sign of its fragility than a sign it is the reigning paradigm—the champion with which to contend.

But at the turn of the twentieth century, this was not the case. Darwin's view was under attack by competing evolutionary theories. Lamarckism, in which changes took place by acquired characteristics—generations of giraffes stretching to eat were thought to pass on elongated necks—was still in vogue more than a hundred years after its conception. Saltationism asserted that new species could be created by one mutation, and orthogenesis held that change was effected only by purpose.

Thus we have this somber description of the fiftieth-anniversary celebration of the publication of *Origin of Species* in 1909 England.

"At an international gathering at Cambridge in July, T. H. Huxley's widow, dignified in black bombazine and bonnet, and the even more ossified widow of Joseph Hooker, Darwin's botanical ally, were paraded out, and many people spoke wistfully of the so-called good old days," writes Hooper. "An undercurrent of sadness permeated the air, for in 1909 Darwinism in England had reached its nadir. These were the dark days that Julian Huxley, T. H. Huxley's grandson, then a student at Oxford, would call 'the eclipse of Darwinism,' when competing theories such as orthogenesis, aristogenesis and Lamarckism were more fashionable than natural selection. For those who felt that the Darwinian jig was almost up, the gaps in Darwinian theory loomed large and many biologists were turning against it."

Darwin's Lasting Contributions

This "virtual funeral" didn't last for long. With the sudden rise of genetic research in the 1940s, Darwin was seen to be vindicated among most biologists. Until then, the material mechanism for evolution was unknown—that is, what was the stuff that carried the change from one generation to another—but with the discovery of DNA, it could be said that evolutionary change meant change that occurred within a physiological genetic code.

This "evolutionary synthesis" was monumental in scientific circles. What's interesting, however, is the impact Darwin's success has had on American culture. Darwin's legacy far exceeds the field of biology; indeed, as Ernst Mayr has said, Darwin changed our zeitgeist.

Darwin's Zeitgeist

Zeitgeist, which stems from the German word "soul" or "spirit" refers to "the spirit of the age." Intentionally general, zeitgeist exists at a cultural level: shaping us not by conscious intention, but by the "ethers" all around us. It is this image that Harvard's evolutionary biologist Ernst Mayr used in his address to the Royal Swedish Academy of Science on receiving their Craaford Prize (along with John Maynard Smith and George C. Williams).

def•i•ni•tion

Zeitgeist is "the general intellectual, moral, and cultural climate of an era." (From Merriam-Webster's Online Dictionary, 2006)

Mayr explains, "Many biological ideas proposed in the past 150 years stood in stark conflict with what everybody assumed to be true. The acceptance of these ideas required an ideological revolution. And no biologist has been responsible for more—and for more drastic—modifications of the average person's worldview than Charles Darwin."

As we shall see, Darwin's science was revolutionary. And as we have seen, Darwin changed the very nature of science itself. Yet what has made his theory so captivating to nonscientists "on the street," is the accessible, immediate effect it has had on the way we view the world.

Darwin's Philosophical Implications

Mayr cites numerous ways that Darwin challenged his own culture and time (and by the looks of the ID conflict, continues to challenge it today). Indeed it may not simply be an odd coincidence that Victorian England and our own contemporary culture should both be in conflict with Darwin. Perhaps the discomfort we feel runs deeper than any particular cultural setting; it may stem from an almost instinctual way we've been given to think about ourselves.

As we've said, Darwin gets rid of "God." Unlike the whole of science until then, the transcendent disappears. Creation is no longer intelligent, and design is only apparent: creator and created are an incidental consequence and expression of evolution.

Thus creation has no teleology. What had been imagined as a purposeful human existence that was headed toward transcendent ends was suddenly the product of accident and circumstance—a journey that just happened to have happened. The purpose-centered worldview popularized by the Greeks, and continued through Galileo and Newton, was suddenly ended with the rise of Darwinism in the middle of the nineteenth century.

In the same way, the deterministic universe, which was conceived to perfection by Newton, was overturned (which opened the way for twentieth-century quantum physics). Variation was random and subject to the accidents of environmental circumstance; so Darwin's science spawned a whole new way of thinking about what now looked like an uncertain future. Chance, the threat to all of us who must be "in control," had shaken this determined universe, and the possibility of purpose.

Bumping Up Against Intelligent Design

As we looked at the three scientific disciplines from the perspective of Intelligent Design, we saw how each implied that human beings were at the center.

The anthropic principle in physics insists that the universe was finely tuned for human life. "Intelligent" DNA implies that chemistry serves human intelligence. And in biology, the "miraculous"—seemingly designed—nature of organic life could be seen as the ultimate product of what God had intended all along.

Still for some, the trouble has been "common descent"—that we all come from a common ancestor. This would have been offensive to the Greeks, to Aquinas, to Galileo, to Descartes, and to Newton; for them, human beings were set apart in the universe to reign supreme. With Darwin, we are faced with the audacious idea that we are not privileged, or special, and are asked to reimagine the nature of our selves and our place in the universe.

The assumption of Intelligent Design extends back to ancient civilizations. Rather than this being the exception to the rule, it is the way we've thought for millennia; until Darwin, what science had presumed is that the universe was woven together by design. Thus Darwin's way of thinking brought the question to the fore as it was never brought before: is there a "designer"? Is there a "God"? Is there a purpose to this life?

The Least You Need to Know

- ◆ Darwin opened the way for a designerless understanding of the physical world.

- ◆ Until Darwin, clergymen were often steeped in the natural sciences.

- ◆ Darwin's voyage to the Galapagos Islands is where his evolutionary journey began.

- ◆ Darwin's theory was fraught with controversy, and the stress of it all purportedly affected his health.

- ◆ Social Darwinism gave rise to such concepts as the superiority and inferiority of races, leading to terrible practices like eugenics.

- ◆ Darwin's theory begs questions of design and purpose.

Chapter 15

Evolution Explained

In This Chapter

- ◆ What is natural selection
- ◆ Overall design
- ◆ Strange—to us—variations
- ◆ What does evolution tell us?

When we hear the word *evolution*, many of us think of the high school text-book illustration of the crouching humanoid caveman progressing through various stages to upright modern man. Others of us might also think of the great nineteenth-century naturalist Charles Darwin. But evolution is much more than the stages of man. And like any theory that has lasted no less than 150 years, evolution has evolved to incorporate generations of scien-tific thinking beyond Darwin.

In fact, evolution simply means "change over time," and refers to more than biology. It is a theory used in chemistry and physics to understand the material world, and as such, is equally concerned with our earliest geologi-cal beginnings. Yet we have seen how, by virtue of the personal way we view organic life, we tend to focus on evolution as a biological theory.

Literally thousands of thinkers have contributed to this sweeping explanation of how organic life developed over hundreds of millions of years. Evolution describes the process of change from the simplest single-cell organisms to the vast organic richness of life that we see all around us today. It explains how biological characteristics are passed on from generation to generation—and how they descended by modification from a common ancestor. The phylogenetic tree (or trees, given the many hypotheses) is an evolutionary map tracing organic life from bacteria to human beings.

The Idea of Natural Selection

According to Darwin, the principal mechanism of evolution is natural selection. This was the great theoretical contribution that was made by Darwin—which persists to this day as the accepted principal engine of evolution. Darwin's book *On the Origin of Species* made natural selection famous (though it was first introduced in a paper coauthored with Alfred Russel Wallace in 1858).

> **Evolutionary Revelations**
>
> Gregor Mendel, abbot of an Augustinian monastery, was an experimental biologist whose obscure, forgotten work brought to life the fading face of evolution. Combining natural selection with the dead monk's experimental theory of heredity began the modern synthesis whereby evolution would emerge again.

Despite the book's early marginal credibility—which followed it into the 1930s—the theory of evolution would ultimately come into its own.

Gregor Mendel proposed that evolution worked by changes in heredity factors. Until Mendel, the question of how changes were passed on was addressed by vague ideas of blending; but with the posthumous discovery of Mendel's work came the start of genetic evolution. By the 1930s, Darwin's theory was exploding in scientific circles to become—in biology as well as in the general culture—the reigning scientific paradigm.

What Is Natural Selection?

Natural selection is the cornerstone of Darwin's theory of evolution. It is the means by which organic species adapt to their environment, and over time works as the catalyst driving evolutionary change.

Until Darwin, species were generally thought to be static creations of God rather than creatures of fluid development as Darwin would come to conclude.

Fact or Faith

> Can it, then, be thought improbable, seeing that variations useful to man have undoubtedly occurred, that other variations useful in some way to each being in the great and complex battle of life, should sometimes occur in the course of thousands of generations? If such do occur, can we doubt (remembering that many more individuals are born than can possibly survive) that individuals having any advantage, however slight, over others, would have the best chance of surviving and of procreating their kind? On the other hand, we may feel sure that any variation in the least degree injurious would be rigidly destroyed. This preservation of favorable variations and the rejection of injurious variations, I call Natural Selection.
>
> —Charles Darwin, *Origin of Species*

Thus natural selection refers to every biological process that causes different traits in organisms which compete to survive and reproduce. When these traits become fixed in organisms, they tend to spread through the population, often resulting in adaptive traits that change the organisms. Because adaptation depends upon the environments in which organisms live, whenever they begin to dwell in different places, they can evolve—even into different species.

Too Many Organisms, Not Enough Food

In the late eighteenth century, Thomas Malthus noticed that human populations always increase faster than the food supply required to sustain them. Darwin applied Malthusian population theory to his own scientific work. In *On the Origin of Species*, he estimated that a pair of slow-breeding elephants—if unimpeded—would produce 19 million elephants in just 750 years.

Beginning from the assumption of scarcity of food, Darwin concluded that organic life was inevitably a battle to survive the material impediments of the world. Given this, he reasoned that organisms endowed with relatively advantageous traits were more likely to prevail, enabling them to reproduce and pass on their advantage. The environments in which they lived—filled with predators and the gnawing need to eat—therefore determined what it meant to be fit and their requirements for adaptation.

So cheetahs, whose lightning speed allows them to escape hungry, large-mouthed lions, evolved to flee their predators—just as lions evolved their vicious teeth to feed their giant appetites. Within the species of the cheetah, it makes natural sense that the slower ones make dinner for the lions, whereas the faster ones live on to reproduce and pass on offspring equal, if not greater, to the task. Generation after generation, as the slower ones fall prey, the faster ones begin to dominate, gradually building this adaptive trait of swiftness into the population.

The moths that adorn high school evolution textbooks provide a celebrated example (however now mired in scientific dispute—see Judith Hooper's *Of Moths and Men*). The industrial revolution in Britain dramatically changed the physical landscape: darkening tree trunks, killing lichen, and giving rise to melanic moths. In the course of 60 years, the once light peppered moths purportedly evolved into dark "morphs," which were favored to elude predatory birds less able to detect them in the woods.

Sexual Selection

A subset of natural selection (selection by environmental factors) is the factor of sexual selection—in short, the capacity to reproduce.

Because reproduction is such a critical variable in the evolution of species, sex plays at least as important a role as it does in our own lives. Indeed, in the light of this focus on sex as a critical evolutionary factor, we might be inclined to examine anew our own sexual behavior.

So combined with environmental viability is this fecundity component. We need only think of human dating rituals—including hours in front of the bathroom mirror—to realize how much they resemble rituals in the animal kingdom. In any case, evolution is critically dependent on the combination of genes, thereby giving mating enormous leverage upon evolutionary outcomes.

Evolutionary Revelations

In the Darwinian view, an extreme example of mating as a critical factor is that of the peacock, with its colorful flamboyance—however unwieldy it might be. In fact, it has been argued that the peacock undermines Darwin's adaptation argument; encumbered as he is, he is therefore less able to fend off his predators. But the Darwinian response is that mating is so critical to favorable reproduction that his feathers confer net advantage as a means to put on one great Friday night show.

Sometimes the line is blurred between environmental adaptation and fecundity. Males' ability to hunt, for example, can be seen as a trait that is attractive to females. And we know sexual fitness can be linked to physical fitness itself—which over generations (in this view) can result in more successfully adapted species.

The Grand Design

A common misunderstanding about evolution is that it is random. As discussed in the following section, randomness is critical to Darwin's pursuit to explain the vast diversity of life. However natural selection is anything but random in the minds of evolution advocates; it is ruthlessly systematic in its approach to effecting physical change.

When difference is established in organisms, natural selection goes to work—searching for traits that confer advantage on an organism's survival. This is the source of its apparent design: the magnificent pursuit of fitness, whereby organisms gradually develop to enable them to realize maximal adaptation. Natural selection is present in every facet of biological life: from semen to freckles, from pollen to pedals, from bacteria to lions in the wild.

Life or Death

Depending on one's viewpoint, natural selection can be seen as destructive or creative—if you're a cheetah in the hungry face of a lion, it may not seem like such a good idea. On the bacterial level, organisms favoring resistance to antibiotics have evolved to fight off being killed in the body, with devastating result.

Evolutionary Revelations

Now it happens that this particular part of the beaver's brain, because of its position in the total wiring diagram, is involved in the beaver's dam-building behavior. Of course, large parts of the brain are involved whenever the beaver builds a dam but, when the … mutation affects this particular part of the brain's wiring diagram, the change has a specific effect on the behavior. It causes the beaver to hold its head higher in the water while swimming with a log in its jaws. Higher, that is, than a beaver without the mutation. This makes it a little less likely that mud, attached to the log, will wash off during the journey. This increases the stickiness of the log, which in turn means that, when the beaver thrusts it into the dam, the log is more likely to stay there. This will tend to apply to all the logs placed by any beaver bearing this particular mutation. The increased stickiness of the logs is a consequence, again a very indirect consequence, of an alteration in the DNA text.

—Richard Dawkins, *The Blind Watchmaker*

Random Mutiny

Natural selection is nothing without initial variation. If there's no variety in a population, selection is powerless to create more-adaptive organisms and ultimately different species. Only when there is variety can natural selection undertake weaving the vast, complex tapestry of life that we see all around us.

The point has been made that natural selection was in trouble in the early 1900s. Competing theories in the air called into question Darwin's elegant mechanism. Specifically, the theory of Jean-Baptiste Lamarck—which was more than 100 years old—continued as a viable explanation for what enabled evolution.

You'll remember that Lamarck believed evolution happened in response to environmental pressures. Seeing increasing complexity throughout the history of living organisms, he explained these changes as a product of demands imposed by their surroundings. Thus the giraffe's elongated neck evolved by stretching for food in trees, to be gradually passed on, thereby expressing the dynamic of acquired variation.

It wasn't until Mendel's rediscovery that attention turned to smaller things. Breeding pea plants, Mendel saw heredity as a product of factors, or alleles, and showed how heredity's building blocks were discrete informational units. Crossing tall plants with short, he found that each next generation was distinctly tall or short—but not a final average of short and tall, as blending would imply.

It was this focus on the dynamics of heredity that ushered in a new era in which the burgeoning genetic research of the 1920s was applied to natural selection. Researchers saw how natural selection could act on genes, producing gradual changes—despite the fact that heredity was a result of distinct, unblended units. Moreover, they found that genetic mutations could produce the very changes that Darwinian evolution required to enable organic variation.

The Odd Mutations

A *mutation*, in the words of Richard Dawkins, is simply a copying error.

Every organism has the capability of replicating its own genetic stuff. But as is the case in all imperfect worlds, such replication isn't always perfect; therefore, over time there is increasing variation by way of these genetic mutations.

Copying errors, during cell division, are largely unpredictable. However, we know they are more likely to happen under certain circumstances: particularly, in the presence of radiation, chemicals, and viruses. This will be important when we talk about the ID position regarding mutation and its argument for what produces evolutionary change—as well as what does not.

def•i•ni•tion

A **mutation** is "a significant and basic alteration; a relatively permanent change in hereditary material involving either a physical change in chromosome relations or a biochemical change in the codons that make up genes."—From Merriam-Webster's Online Dictionary, 2006

Errors Don't Mean Mistakes

But imperfect replication need not be considered a negative thing in the world of evolution. It is the critical stuff of variation, without which evolution couldn't happen.

In Dawkins's way of thinking, mutation, in fact, deserves significant credit: "… instead of their being a uniform population of identical replicators, we shall have a mixed population. Probably many of the products of erratic copying will be found to have lost the property of self-replication, while being different from the parent in some other respect. So we shall have copies of errors being duplicated in the population."

What Do Mutations Do?

Returning to our Mendelian geneticists who were working in the 1920s, genetic mutations were believed to produce small changes in organisms' traits. Rather than a series of dramatic leaps, evolution could be seen as gradual; given enough time, or reproductive generations, mutations could alter populations. As natural selection worked on these genetic changes, the ones conferring greater fitness would be retained, whereas genes detrimental to survival would be more likely to disappear.

Definitions of the term *random* are hotly disputed. Some, like Dawkins, resist the word in the name of natural selection. Proponents of Intelligent Design use it negatively. Mutations represent unpredictable deviations from genetic norms, leaving the future to chance (and where we are going, to blind materialism). ID's need for direction, perhaps a hand of God, is dismissed by evolutionists in favor of embracing the uncertainties implied by an exclusively material existence.

In any case, the apparent purpose or design one might see in the universe is, according to most evolutionists, finally an illusion. The creation of species from a common ancestor, and all subsequent organismic change, are a result of diversity-producing mutations, which are sorted through by natural selection. If the order that one sees is purposeful—it is not to some "intelligent" end—it reflects this magnificent machine of evolution interacting with genetic imperfections.

That said, when we look at neo-Darwinism, we shall have to ask the question: is it only the proponents of ID who resist the idea of randomness? One might argue Richard Dawkins comes remarkably close to the Intelligent Design position that there is some kind of systematic formula by which life takes on purposeful direction. It is important, in this very often subtle debate, to look beneath such words as "random"—to see if they are pointing to the meanings we assume, or if they are in some sense, deceptive.

What Does Evolution Explain?

Watson and Wilson called Darwin the greatest figure in history. Consequently, they might also call evolution the greatest theory in history.

For many people—both inside and outside of the overall scientific community—the question isn't "What does it explain?" In these halcyon days of evolution, it may be rather, "What doesn't it explain?"

Evolution gives credence to the fossil record that traces the development of life, helping us to understand how species developed and the diversity of organic life. It describes the locomotive of organic change—in terms of natural selection—and how organisms survive and reproduce in response to their environments. So it strives to explain from whence we come and perhaps where we are going, shaping not only our physical existence, but a sense of the meaning of our lives.

Although biology is just one field among many other scientific disciplines—chemistry, physics, geology, all vying for some all-explaining theory—given its connection with human life itself, it is bound to bear the brunt of criticism, engendering anger and resentment among those on either side of the design debate. No doubt this is why evolutionists and ID advocates tend not to get along: for thousands of years, such all-explaining theories presumed the existence of God. If Darwin took God out of the equation in unprecedented ways, it was his adherents who came to perceive evolution as final proof there is no God.

Evolving Populations

It is easy to lose the forest for the trees about evolutionary theory. As we have said, evolution isn't simply about biology or chemistry; it includes geology, botany, physics, paleontology, and a host of other disciplines. This is most apparent when looking at the effects of evolution on populations of species that change as a result of such seemingly unrelated factors as the weather.

Fact or Faith _____

The migrations of species, their population size, and indeed, random fluctuations all contribute to the nature of their evolution, and finally what they come to be. So life becomes a drama of chance and accidents, followed by the natural selection of those genetic traits which best confer survival upon evolving populations. Thus the animal kingdom, including human beings, may—to many observers—no longer be seen as static creatures that were rendered by God's hand in the Garden of Eden.

Because natural selection is a process of local environmental adaptation, there is never any overall "purpose" or "direction" to organic change. Theistic claims of Intelligent Design—by which God directs the miracle of life—are for most Darwinians discarded by the inherent randomness of evolution. Purpose, as presumed by proponents of ID, necessarily disappears, thereby dethroning the preeminence of human beings as the reason for creation.

The Evolution of Man

Of course central to the ID debate is the evolution of human beings. It should be remembered that many ID advocates have no problem with this idea. However, given the anthropic arguments for a finely-tuned universe, the place of human beings, however evolved, asks as many questions as it answers.

The human implications of evolution were clear in Darwin's lifetime. Although he didn't specifically address the issue in *On the Origin of Species*, he did say, "… light will be thrown on the origin of man and his history."

By the time he published *The Descent of Man* 20 years later, it was clear that the idea of common ancestry, or homology, was presumed. The idea was controversial from the start, even with Alfred Russel Wallace—Darwin's cofounder of natural selection in 1858.

Men from Monkeys?

In the nineteenth century, it was commonly believed—at least in scientific circles— that human beings shared common ancestry with gorillas and chimpanzees. But in 1924, Australopithecus Africanus was discovered in South Africa, with a small, rounded brain, unlike the chimpanzee, resembling that of a human. Although it took 20 years to be accepted as a transitional ancestor, it is now believed to be part of the genus that preceded genus Homo sapien.

Homo sapiens are the only remaining species of the homo genus. Beginning about 2.4 million years ago, Homo habilis emerged, followed by six other Homo species, until our own: Homo sapien. Although such classification is complex, subject to argument, and contested even among evolutionists, it is generally agreed that modern man appeared several hundred thousand years ago.

For evolutionists, the fossil record indicates the gradual evolution of man. For some it may not only vindicate Darwin, but the power of evolutionary thought to explain the way that human beings came to be, and their place in the universe. Indeed, it's not uncommon for evolutionists to leap to the conclusion that this special place (which is asserted by Intelligent Design) is, in fact, no special place at all.

Where Do We Go Now?

It is clear that the argument for Intelligent Design is philosophically invested in human beings playing a critical part in the design of the universe. ID advocates, who believe in evolution, still see the evolutionary process as designed by an intelligent agent, which inherently bears transcendent meaning. As we have said, this is not an unprecedented scientific assumption; the greatest scientists over several thousand years presumed Intelligent Design.

In the same way, we shall see how neo-Darwinists take an approach similar to ID. Using science to bolster their position about God, most adhere to atheism as a consequence of their inquiry into the material universe. In either case, it is clear that these positions are themselves philosophical leaps of faith: one supposes God manifestly exists, whereas the other contends God does not exist.

So we turn to neo-Darwinism and then to Intelligent Design—in our quest to get to the bottom of the Intelligent Design debate. Having laid the groundwork, now we are ready to explore the philosophical issues that surround—and one might argue, permeate—the very nature of science itself. Thus we turn to neo-Darwinism.

The Least You Need to Know

- ◆ The principal mechanism of evolution is natural selection.

- ◆ Natural selection is the means by which species adapt to their environment and over time works as the catalyst driving evolutionary change.

- ◆ According to Darwin, organic life is in a constant battle to survive the material impediments of the world.

- ◆ Sexual superiority plays an important role in natural selection.

- ◆ Proponents of Intelligent Design are not necessarily opponents of evolution.

Chapter 16

Evolving Atheism

In This Chapter

- ◆ What the new Darwinists say
- ◆ The "mechanical" Dr. Wilson
- ◆ A self-centered gene
- ◆ The "radical materialism" of Dennett

Neo-Darwinism is a school of thought that began with the modern synthesis of Darwin's theory of natural selection with Mendel's genetic discoveries which he made by breeding plants. Now—for the first time—there was a mechanism for natural selection that could be scientifically explained, thus vindicating Darwin's theory that had been in decline into the 1930s.

Mendel's idea of heredity factors paved the way for genetic research. The units of evolution could be understood in terms of genes, whose competition for survival precipitated evolutionary change. For neo-Darwinists, the discovery of genes was the key that unlocked Darwin's door, shedding light upon the specific process that explained natural selection.

Assertions of the New Darwinists

It is understandable how genetic research, which was exploding onto the scene—offering up the missing pieces that were needed to complete the evolutionary puzzle—would capture biologists' imaginations. Those imaginations, after all, were burning with the gnawing suspicion that—however incomplete their understanding remained—natural selection was indeed at work.

When the answer emerged in the form of genes and the DNA molecule, it seemed to many that natural selection was able to explain everything.

Refuting the Existence of God?

Not only was natural selection able to successfully describe biological processes; for many it called into question human meaning, and indeed, the necessity of God.

Remember that Darwin had insisted on natural selection being self-sufficient—that is, requiring no transcendent, guiding hand to direct life to some predetermined end. Many neo-Darwinists take Darwin's insistence to mean more than that God isn't required. By asserting that all of life can be materially explained, they assert that God does not exist.

ID Is Not Science

An apparent contradiction in the ID debate is the neo-Darwinist assertion that Intelligent Design isn't really science because it infers a transcendent agent. If this is the case, neo-Darwinism is guilty of nonscientific claims insofar as it departs its own material world to contend that there is no God. As we shall see, however, the definition of science is at the very heart of the conflict—being used as a weapon on either side of this politically charged debate.

So we look at three leading voices of neo-Darwinism who have in various ways utilized science to challenge belief in God:

   ◆ Evolutionary biologist Edward Osborne Wilson, who represents the most open point of view, resisting belief in God on the grounds that it lies beyond scientific confirmation

◆ Richard Dawkins, Professor of Public Understanding of Science at Oxford, who applauds Darwin for allowing his adherents to become intellectually fulfilled atheists

◆ Daniel Dennett, a philosopher at Tufts University, who has recently published a book about how religious faith is no more than a spell that he believes has to be broken

Dr. Wilson's Mechanical Men

Edward Osborne Wilson is Professor Emeritus of Biology at Harvard. The world's foremost authority on species of ants, he used his study of ant behavior to try to understand the nature of human life in a social context. His book *Sociobiology: The New Synthesis* (1975) and a speech he gave which met with a pitcher of water over his head stirred some controversy.

Asserting all behavior is reducible to a person's genetic constitution—from violence, to sexism, to patriotism, to the ability to compromise—Wilson argued that behavior is, at the bottom, a biological phenomenon, and one can do little to alter his own genetic determination. Even music is here perceived as a genetic adaptation, conferring physical advantage on artful populations that are fortified to live another day.

Colleagues, several from Harvard, condemned his theory in a letter to the *New York Times*.

Sociobiology

As we have seen, natural selection seeks biological survival. Until genetic research, certain phenomena were hard to explain by evolution. For example, how is it that parents should willingly give up their lives on behalf of their children—or, in the case of Wilson's colonies of ants, that worker ants die to protect their nests?

But sociobiology looks beyond the survival of individual organisms to the overall social context in which the organisms find themselves. Wilson asserts that because worker ants are all genetically related, they are therefore motivated to protect the common genes that as a colony they share together. Thus survival isn't simply a function of the fate of individual organisms; it is a function of the gene pool of the colony—just as it is in human families.

Genetic survival has newly defined Darwin's concept of natural selection, instinctually seething from an organism's pores to determine its social behavior. Though Wilson concedes that environmental factors contribute to human behavior, he admits to his own Darwinian, materialist bias. In *Sociobiology: The New Synthesis*, he writes, "The central idea of the philosophy of behaviorism, that behavior and the mind have an entirely materialist basis subject to experimental analysis, is fundamentally sound …. The learning potential of each species appears to be fully programmed by the structure of its brain, the sequence of release of its hormones, and ultimately, its genes."

Man as a Machine

Like it or not, the conclusion one might draw is that human beings are machines. Responding to their own biochemistry, which interacts with the environment, humans

Fact or Faith

For the purposes of the Intelligent Design debate, an implication of Edward Osborne Wilson's work is that the souls of human beings are themselves an illusion—perhaps for better adaptation.

are not free to act on their own as a function of self-conscious, free will. So Wilson entertains the possibility that human freedom is no more than an illusion: even less than a machine, we are cogs in a machine which is assembled by the larger population.

In this light (that man is a cog, and souls are illusions), one could argue that faith in God confers better chances of survival under grimmest circumstances, where a realistic view would be likely to render despair.

Of course, this is not the same as the belief in the soul as a unique, transcendent, vital life, which is borne in human beings, and indeed, which testifies to the transcendent hand of its creator.

In this regard, it may be neo-Darwinists are right: it's science or religion. Their materialist view leaves little room for human freedom, human choice, or transcendent human meaning. Even falling in love is here reducible to neurons firing in the brain—like a host of other needs that interact to comprise maximal genetic survival.

The implications are clear—and theological. Within this perspective, there is no God or purpose beyond a material process. Our sense of self (which we attribute to our human souls) and the sense of God (we may experience in prayer) are nothing more than genes struggling to survive and replicate into another generation.

Neo-Darwinist Metaphysics

Many neo-Darwinist claims are theological in nature; and as such, they might be most accurately assessed as competing articles of faith. Indeed, their implicit belief in science as the ultimate avenue to truth is itself unprovable and can't be verified beyond its own prior assumptions. In this way, tautology—a criticism waged against proponents of Intelligent Design—may also describe the problem which plagues this new mythology of science.

Wilson makes reference to the mythology of scientific materialism—a term that seems to tentatively approach the realm of the religious. "What I am suggesting, in the end, is that the evolutionary epic is probably the best myth we will ever have," he writes. "It can be adjusted until it comes as close to the truth as the human mind is constructed to judge the truth."

Fact or Faith _____

Science offers the boldest metaphysics of the age: the faith that if we dream, press to discover, explain, and dream again, thereby plunging repeatedly into new terrain, the world will somehow become clearer and we will grasp the true strangeness of the universe, and the strangeness will all prove to be connected and make sense.

—Edward Osborne Wilson, *Sociobiology: The New Synthesis*

Herein lies a problem for neo-Darwinists. For if the mind is a product of evolution, then such truth is qualified by its source. If the brain is a product of evolution, how can it know truth beyond itself? Thus religion as evolution's adaptation applies to evolution, too; and so we are left with the possibility that no one truth is better than another.

Having It Both Ways

Yet Wilson considers himself a deist who considers the possibility that a creative force may have had a hand in beginning the universe. Still his sociobiology leaves little room for transcendent dimensions, and a lot of room for the material as the sole medium for human existence. If there is a larger reality for Wilson, it is incarnated by the gene—whose self-replication is an undirected, purposeless, self-perpetuating mechanism.

"In a Darwinian sense the organism does not live for itself," Wilson writes. "Its primary function is not even to reproduce other organisms; it reproduces genes, and it serves as their temporary carrier … Samuel Butler's famous aphorism, that the chicken is only the egg's way of making another egg, has been modernized: the organism is only DNA's way of making more DNA."

Dawkins and *The Selfish Gene*

Richard Dawkins is for many the spokesperson for neo-Darwinism in our time. Author of *The Selfish Gene*, which since the 1970s has sold more than a million copies, his gift for communicating scientific ideas has contributed to making science famous. It is arguable that among the leading Darwinists, Dawkins is the theologian, often indicating that a natural conclusion of evolution is there is no God.

You will remember that a centerpiece of nineteenth-century British theology was the argument that God was manifest in the design of the natural world.

Philosopher William Paley asserted that God's evidence resided in the stuff: in the delicate balances, in the physical laws, and in the structures of the universe. He likened the world to a watch that was suddenly found in an open field—which one could only conclude had been designed rather than naturally created.

Blinded By Science

Dawkins tips his hat to Paley on the way to dismissing his argument, saying Paley's book is to be admired but is "wrong, gloriously and utterly wrong."

This argument for God as the natural designer is utterly opposed by Dawkins. The title of his book *The Blind Watchmaker* probably says it all. If there is a creator, the creator is blind and devoid of teleology or purpose; his creator, in fact, is simply evolution acting in a godless world.

Dawkins says that Darwin successfully explained how Paley's design is only apparent. Until Darwin, Dawkins patronizes Paley's argument, the argument made sense. In other words, Paley and Dawkins agree about the evident complexity of life; the trouble, Dawkins says, is not the evidence—but any theistic explanation.

Selfish Genes

Dawkins's ability to attract attention is seen in his title *The Selfish Gene*. In fact, it is ironic that he seemingly attributes material genes with motives—given his rejection of the natural world as having intention or purpose. But beyond simply being provocative, Dawkins wants to focus on genes as the fundamental element that moves the evolutionary mechanism ahead.

Dawkins contends that every gene naturally seeks its own interest. Genes that are passed on are the ones that successfully act to benefit themselves. Although most of the time genes' interests are the same as those of their host organisms, sometimes genes' interests are put before their hosts'—to the detriment of the organism.

An example of the first is a gene that protects its host organism against disease. An example of the second (that is, genes which conflict with the interest of the host organism) is a gene that drives a male spider's mating instinct at the cost of being eaten by the female. In this second case, however, one sees how the organism isn't the principal unit: by successful reproduction, the gene's life is replicated, which for Dawkins, is the only game in town.

For Dawkins, genes build survival machines in the form of larger organisms. Yet genetic evolution not only answers questions to be answered inside organisms; it also answers questions in the social arena—at the level of human behavior. For example, one might see the sacrifice a mother exercises on behalf of her child as an expression of the need, motivated by her genes, to perpetuate her own genetic pool.

Memetics

Speaking of social implications inherent in Dawkins's way of thinking, Dawkins coined the term *meme* to extend genetic theory into the cultural arena. Although memetics is controversial in the minds of many evolutionists, not unlike Wilson, Dawkins perceives social behavior in evolutionary terms.

In *The Selfish Gene*, Dawkins explains what a meme is: "Examples of memes are tunes, ideas, catch-phrases, clothes fashions, ways of making pots or of building arches. Just as genes propagate themselves in the gene pool by leaping from body to body via sperms or eggs, so memes propagate themselves in the meme pool by leaping from brain to brain via a process which, in the broad sense, can be called imitation."

Dawkins is contending what many of us know: that human culture has a life of its own—and habits, practices, and tunes are passed on from group to group, and person to person. But believing evolution is too powerful a theory to be confined to mere biology, Dawkins sees the theory's own powerful evolution into the social arena. However vague may be the concept, memes are cultural genes that behave for their own survival and in the process also act to maximize survival of the larger population.

From relatively noncontroversial examples of memetic evolution, Dawkins invariably moves to the subject of God and theology. So the belief in life after death is a meme one might see as conferring advantage on cultures that otherwise might be compromised by the prospect of death. Yet Dawkins is clear to assert this is genetic adaptation to the end of survival, and never what he would see as an authentic experience of the transcendent.

Dawkins's Atheism

For years Dawkins generally escaped criticism for his atheistic conclusions. Recently, however, there has been a growing recognition of the problem Dawkins's stand implies.

The atheism itself is not the problem—we've seen that many, perhaps most, scientists are atheists—it is rather that the claims which Dawkins makes are made under the guise of science. Beyond the problem that such theological claims lie beyond Dawkins's expertise, Darwin's own demand for exclusively material explanations is thereby undermined.

> **Evolutionary Revelations**
>
> In November 2000, Dawkins gave a lecture titled "Evolution: The Greatest Show on Earth, the Only Game in Town."

However Dawkins strives to translate his theology into scientific terms, critics perceive a theological agenda reminiscent of creationism. As we shall see, this is how the fight is increasingly being fought—rather than in scientific terms or in the light of the nature of human knowledge.

The following excerpt from Dawkins's book is an example of religion being used as a whipping boy for the apparent purpose of a memetic explanation which seems unduly complicated:

"Consider the idea of God. We do not know how it arose in the meme pool. Probably it originated many times by independent 'mutation.' In any case, it is very old indeed. How does it replicate itself? By the spoken and written word, aided by the great music and great art. Why does it have such high survival value? … The survival value of the god meme in the meme pool results from its great psychological appeal. It provides a superficially plausible answer to deep and troubling questions about existence. It suggests that injustices in this world may be rectified in the next. The 'everlasting arms' hold out a cushion against our inadequacies, which, like the doctor's placebo, is nonetheless effective for being imaginary. These are some of the reasons why the idea of God is copied so readily by successive generations of individual brains. God exists, if only in the form of meme with high survival value, or effective power, in the environment provided by human culture."

Dawkin's Purpose

In his commentary, titled, "A Journalist's Struggle to Remain on the Sidelines in the Evolution Wars," science writer Dick Teresi questions the neo-Darwinist, in particular, Richard Dawkins's claim that evolution is ultimately "random." He notes Dawkins's insistence on progressive improvement, and the way the neo-Darwinists tend to fill evolution with "meaning, order, causality." Of course the philosophical problem is that if natural selection so effectively and specifically determines evolutionary outcomes, where is the randomness?

It's in the mutations, neo-Darwinists claim! Still, life is getting better and better. And if it's getting better, who determines what is better? And if better, then must there be a best?

The point is that "purpose," or "teleology," is so deeply written into our beings that even those who make it their life to reject purpose seem at a loss to totally reject it. One could argue that Dawkins has simply interchanged "natural selection" with "God," and that the guiding hand of natural selection looks like God's hand in the Sistine Chapel. At this point, the question becomes theological—or at least, philosophical—and Dawkins violates the very criticism he's waged against Intelligent Design.

Dennett's Radical Materialism

Daniel Dennett, a leading Darwinian philosopher, is considered an ally of Dawkins. The recipient of numerous academic honors, he has risen to prominence by such controversial books as *Darwin's Dangerous Idea* and (on religion) *Breaking the Spell*. Though not a biologist, Dennett's focus of study is evolution and human consciousness, and so he has figured largely in the Intelligent Design debate.

Dennett is a radical materialist: all can be reduced to mechanisms. Faith, hope, love, dreams, art, and memory can be reduced to a material process. His belief in evolution, whatever the scientific evidence might indicate, has rendered him a leading Darwinian, proponent and critic of Intelligent Design.

Universal Acid

For Dennett, Darwinism has the potential to explain everything. Although Dennett's perception of religious faith wasn't shared by Darwin himself—nor for that matter, Dennett's distaste for anything remotely religious—Dennett appears to deduce atheism from Darwin's materialism, and so, in a way similar to Dawkins, is an outspoken atheist. In fact, it isn't simply religion at which Dennett takes specific aim, but any philosophical idea that competes with Darwinian explanation.

So Dennett asserts—in *Darwin's Dangerous Idea*—that Darwinism is a universal acid that eats through just about every traditional concept and leaves, in its wake a revolutionized worldview.

No Mere Biological Theory

Thus for Dennett, Darwinism isn't simply to be seen as a biological theory; it is a grand philosophical view of the world, and indeed, of human existence. The fierceness with which the Intelligent Design debate is being fought, testifies to Darwinism's implications beyond the bounds of scientific discourse.

Dennett writes, "Darwin's idea had been born as an answer to questions in biology, but it threatened to leak out, offering answers—welcome or not—to questions in cosmology (going in one direction) and psychology (going in the other direction). If (biological design) could be a mindless, algorithmic process of evolution, why couldn't that whole process itself be the product of evolution, and so forth, all the way down? And if mindless evolution could account for the breathtakingly clever artifacts of the biosphere, how could the products of our own 'real' minds be exempt from an evolutionary explanation? Darwin's idea thus also threatened to spread all the way up, dissolving the illusion of our own authorship, our own divine spark of creativity and understanding."

Looking on the Bright Side

Dennett has managed to position himself in the political fray. In an op-ed piece published in the *New York Times* titled "The Bright Stuff," he allied himself with other atheists, to whom he gave the name of "Brights." For many "non-Bright" theists, the message was clear, naturally inciting resentment, and making it difficult to conduct a civil discourse, or constructively address the disagreement.

We will see the ways such fervor has impeded the debate on both sides of the political fence. Dennett says that atheists simply want "to be treated with the same respect accorded to Baptists and Hindus and Catholics, no more and no less." Yet it is difficult to see how referring to oneself as a Bright, with its implications about others, might contribute to the same genuine respect for which Dennett himself is asking.

When we get to the chapter that deals with politics, we may begin to see a source of the problem. A clear majority of Americans believe that God created the world, whereas a majority of scientists reject divine creation for evolutionary explanations.

This disparity may mark an educational—and probably a socioeconomic—split as much as it may signal where sympathies lie on a purely political spectrum.

If there is to be a constructive engagement of ideas (rather than another food fight), there will have to be a newborn respect exhibited from both sides. It would seem that academics, who presume their liberalism is proof of their respect for others, may need to dig deeper than political positions to listen to the other's concerns. Likewise, "the other's" view of ivory tower academics who care little for unwashed nonacademics must give way to understanding they are fellow human beings on a common journey toward the truth.

The Least You Need to Know

- Neo-Darwinism began with Mendel's modern synthesis, which formed an explanation for natural selection.

- The nonrandom nature of natural selection begs the question of "purpose," even despite its atheistic implications.

- E. O. Wilson, Richard Dawkins, and Daniel Dennett are neo-Darwinists who, through science, have challenged the belief in God.

- Wilson resists belief in God as lacking scientific confirmation.

- Dawkins applauds Darwin for creating a scientific environment that respects atheism.

- Dennett views religious faith as a spell that should be broken.

Part 6 "Intelligent" Meaning of Life

Contrary to what many in the general public believe, proponents of Intelligent Design do not reject all Darwinian theory: much of evolution is in fact embraced by ID proponents, who assert that not all apparent "design" could have gradually evolved. ID proponents believe there is a more adequate explanation than that which has been adopted by present-day evolutionists.

Gospel of Darwin, According to ID

In This Chapter

- ◆ Choices, chances, and copy errors
- ◆ How selection works
- ◆ The science of evolution

Charles Darwin wrote, "If numerous species, belonging to the same genera or families, have really started into life all at once, the fact would be fatal to the theory of descent with slow modification through natural selection."

This qualification by Darwin concerning the credibility of evolution was not likely made to question the theory's scientific content. In all likelihood, it was made to show it was falsifiable; and that put to the test, evolution could meet all competing theories. Although there are some ID marksmen who believe in (at least certain forms of) evolution, their sights, which appeared to be aimed toward evolution, may be trained on Darwin himself.

A corollary, which you may recall from Chapter 13, is present in Michael Behe's appeal to Darwin's own assertion, which said, "If it could be demonstrated that any complex organ existed which could not possibly have been formed by numerous, successive, slight modifications, my [Darwin's] theory would absolutely break down."

As we have said, there is something personal about biology and evolution that somehow doesn't seem to be present in the fields of chemistry and physics. Where we all come from, and where we are going, are questions of human meaning, and so beg for religious and philosophical answers. Indeed, of all the traditional scientists, biologists appear to be most eager to speculate (by way of their scientific view) that God does not exist.

So we move to Intelligent Design as a competing school of thought that attempts to challenge evolution upon scientific grounds. In Chapter 18, we look more closely at some of ID's underpinnings—that is, its apparent monotheistic Christian presuppositions. For now, however, our focus is on ID's critique of evolution's dynamics: random mutation, natural selection, and their scientific credibility.

Randomly Copying Errors

What Darwin accomplished in *Origin of Species* was clearly no small feat. Evolution is an overarching explanation for every organism: from bacteria to coral, from flies to chimpanzees, from mushrooms to roses, to us. In one sweeping description, Darwin unified our vision of the whole of life as a singular function of two dynamics: mutation and natural selection.

You may remember that for evolution to occur by way of natural selection, it must first have variation, to select from several relevant competing genes. So the question would become: how did we get variation in the first place? How can selection effect its change if there are no choices to be made?

Ah, the key word, *choices.*

The evolutionary answer, now translated into modern genetic terms, is that random mutation creates diversity, and so the stuff on which selection operates. But for advocates of Intelligent Design, this is a dubious explanation. At best it's an unfounded biochemical assumption—which has never been adequately tested.

Too Many Chances

In his book, *The Science of God*, author Gerald Schroeder notes that humans have about 70,000 genes, which in turn organize about 70 million amino acids into organic structures. Because this makes 30 trillion cells in a healthy human being, he ponders the question of whether random processes could generate such miraculous structure. Is it really reasonable to assume that these highly complex machines could have come about by random chance—without the help of some intelligent agent?

An analogy has often been made to the proverbial room full of typing monkeys who are randomly banging on typewriters to the end of creating a meaningful sentence. Of course, the reason these monkeys never achieve a sentence more than several words long is that the number of meaningless letter combinations far exceeds meaningful combinations.

> **Evolutionary Revelations**
>
> The typing monkeys analogy is often attributed to nineteenth-century biologist Thomas "Darwin's Bulldog" Huxley.

In fact, to get to any meaningful sentence is, in Schroeder's mind, no small task. "With one hundred letters in a sentence, there are 26,100 combinations of those letters," he writes in *The Science of God.* "If one were written on each fundamental particle in the entire universe, it would take literally billions of billions of universes to complete the task of printing out the 'text.'"

A greater problem than evolution's apparent randomness may be how it all began in the first place. As we saw in Chapter 12, ID proponents question how organisms first emerged from the gases. The best mainstream answer appears to be random chemical generation, by which chemicals in the primeval soup coupled to produce organic life.

Hoyle's 747

Yet the British astronomer Sir Fredrick Hoyle finds this scenario unlikely. He said the chances of this happening were equal to the chances of a Boeing 747 being randomly assembled by a tornado blowing through a junkyard. It appears that this problem may have inspired three University of Michigan chemical engineers to recently call for more open debate about the theory of evolution.

Thus is random chance unconvincing to Intelligent Design advocates—and indeed, may be unconvincing to most Americans who believe the universe was designed. As politically charged as Intelligent Design has become, it is as old as ancient Greece—indeed, even neo-Darwinists such as Richard Dawkins admit to the appearance of design. Within the ID debate, the problem isn't how the evidence appears; it is whether it happened by material process or by the hand of some transcendent agent.

"Besides, Mutations Are Usually Bad"

One of the basic assumptions about random mutation in evolutionary theory is that mutations are good because they lay the groundwork for diversity of life. By replication errors, chemical exposures, and other causes of mutation, the stage is set for DNA changes that can eventuate in different species. Rather than gross images of fruit flies with legs growing out of their heads, in the neo-Darwinist view mutations can result in organisms fitter for survival.

> **Evolutionary Revelations**
>
> Laboratory scientists have indeed produced such a gross mutation as the fruit fly called antennapedia. Although this is an intentionally chosen example of random genetic mutation, for the purpose of explaining the ID position, it illustrates the point.

In the ID camp, mutations are generally bad—which is to say, destructive—and are most commonly seen in such devastating illnesses as cancer and cystic fibrosis.

It should be noted that one neo-Darwinist response is that the destructive mutations get the press; because medical research is focused on health, we are most aware of "bad" mutations. Moreover, minute genetic changes that confer unseen advantage on organisms in the long run rarely see the light of day—and yet (Darwinists claim) they are nevertheless at work. To which ID proponents say there is no evidence of such evolution or speciation; rather, they believe, as Michael Behe has said, it generally amounts to hand waving.

Natural Selection

It was natural selection that was the centerpiece of Darwin's revolutionary theory. If random mutation has been increasingly stressed in the light of genetic research, natural selection, as it flowed from Darwin's pen, has been adopted pretty much in tact. Survival of the most fit remains the engine of genetic natural selection, whereby organisms improve in response to their need to survive in their environments.

The apparent design of which Richard Dawkins speaks is created by natural selection. As organisms develop to meet survival needs, they manifest required complexity: whether a cheetah's swiftest legs, or a lion's razor teeth, or a peacock's grand and ostentatious tail. When a mutation occurs, selection can choose it as a function of survival improvement, and so the vast diversity of life the world enjoys naturally appears to be designed.

For ID proponents, such design could not derive from natural selection alone. The complexity inherent in organic life cannot be so explained by evolution. Simply put, there are too many holes and contradictions in both the theory and the evidence.

Dawkins's Monkeys

One of the pivotal ID arguments against Darwinian natural selection is that it can't produce the complexity inherent in living organisms. If natural selection works on random mutations to produce, for example, an eye—favoring those genes that allow the eye to function to the advantage of the host organism—we should be able to calculate the number of required mutations in our DNA, and so the capacity of evolution to produce such a complex machine. In the evolutionary scenario, Gerald Schroeder believes the chances of this are minute.

> **Blinded By Science** _____
>
> If the evolutionary model that we choose is such that each of the thousand steps in our hypothetical mutation must be in sequence and any out-of-order mutation is fatal, the number of trials required in the process is 4 (to the 1,000th power) or in the usual decimal notation, 10 (to the 600th power). That is a one with six hundred zeros after it! And that is just to get the information of an eye to the brain. We didn't start the processing of that information by the brain.
>
> —Gerald Schroeder, *The Science of God*

But Schroeder acknowledges these particular assumptions are overly restrictive—that erroneous mutations may not kill the chances of evolving an eye. On the other extreme, too optimistic a model evolves an eye with no problem. If the mutations could occur in any order, and incorrect mutations were never fatal—and if the correct mutations were always preserved and never mutated away—the 25 percent likelihood that each (nucleotide) mutation would be successful means that after only 10 generations, the evolution of an eye would be achieved.

Monkeys Typing Shakespeare

Noting how different this is from the model where each mutation has to happen in order, Schroeder says this highly unrealistic assumption is the one used by Dawkins. In *The Blind Watchmaker*, Dawkins describes how natural selection acts like the proverbial monkeys who, given enough time, type all of Shakespeare. In this case, he takes a line from Hamlet—"Methinks it is like a weasel"—explaining that the chances of a monkey typing it in one fell swoop would take millions of years.

> ### Evolutionary Revelations
>
> A 2003 study in Plymouth, England, of six monkeys in a lab with a computer for one month did not produce anything but a mess.
>
> According to the Associated Press, the primates "attacked the machine and failed to produce a single word …. 'They pressed a lot of S's,' researcher Mike Phillips said Friday [May 9, 2003]. 'Obviously, English isn't their first language.' … At first, said Phillips, 'the lead male got a stone and started bashing the hell out of it.' … 'Another thing they were interested in was in defecating and urinating all over the keyboard,' added Phillips. … Eventually, monkeys Elmo, Gum, Heather, Holly, Mistletoe and Rowan produced five pages of text, composed primarily of the letter S. Later, the letters A, J, L and M crept in."
>
> Not one word, however.

This "single-step selection," Dawkins says, is different from "cumulative selection." Cumulative selection does not start from scratch, but builds upon past improvements—thereby allowing for a vast complexity that, by single steps, would be impossible. Natural selection, Dawkins contends, is an example of cumulative selection; thus he answers critics who contend natural selection couldn't produce organic life.

The problem, Schroeder points out, is that Dawkins proves the ID claim he seeks to undermine: that natural selection alone cannot explain the complexity of organic life. Intelligent Design suggests there has to be an agent that explains life's design—in the case of evolution, an agent who might guide the biological process. ID's contention that natural selection must know the end to which life is inclined is ironically confirmed by Dawkins's program, whose goal is, "Methinks it is like a weasel."

Schroeder asserts, "Dawkins's model had a known goal and worked toward that goal, knowing which letters it wanted in each of the 28 slots. It is an ideal demonstration of directed evolution." If natural selection is truly at work, ID advocates might say, it is no more undirected than the predetermined purpose of reproducing a sentence from Hamlet.

The Problem of Purpose

As we have seen, evolutionists take pains to say that purpose is absent from their theory. As systematic as natural selection may be, it creates only local adaptation. There is no overriding plan, or design, by which organic life progresses.

This Darwinian insistence on a lack of plan seems critical for two essential reasons. The first has to do with the nature of the science—and of how evolution works. The second is more philosophical in nature, having to do with extra-scientific questions: specifically, how one finds existential meaning without the interference of God.

With regard to the first, we have seen how survival of the fittest drives social selection. Offensive as it seems to perceive life as a process of pursuing one's own self-interest, most of us have known how effective it is in effecting survival and success. It is evident in genetics, too—in finches' beaks, bacteria, and human beings—all in response to a very local fight to survive in their environments.

Thus the process of adaptation is an entirely material dynamic. There is no need to explain any teleology beyond the material itself. Evolution is self-sufficient, and for the first time in the history of science departs from several thousand years of transcendent assumptions to bask in its autonomous success.

With regard to the second, philosophical reason for insisting on a lack of purpose, acknowledging a guide, or a transcendent design, defies the current culture of science.

It should also be noted that in the light of the modern scientific method, the introduction of God into the physical world seems like unnecessary baggage. One of the tenets of science is parsimony: being spare in scientific explanation. Beyond the scientific problem of making an inference of God (with which we will soon deal) is that of defying the nature of good science—which is free of superfluous ideas.

Fact or Faith

Darwin effectively redefined the nature of science—following the Greeks, Galileo, and Newton—to finally squeeze God from the equation. Science is now perceived to operate as a singularly material process; and for many scientists, it has naturally led to a philosophy of atheism.

The Tree of Life

The evolutionary tree, which is popularly drawn to map Darwin's "descent with modification," has in recent years been dramatically revised to accommodate the Cambrian explosion. You may remember this event took place more than 500 million years ago, producing all the body plans for living animals that have lived until the present day. The sudden explosion was so dramatic that paleontologists speculate new animal types may have been produced in just several thousand years.

Fact or Faith _____

Darwin's gradual descent with modification is challenged by the fossil record. There are clearly solutions—like Gould's and Eldridge's "punctuated equilibrium." Nevertheless, for ID advocates, it is as though "someone knew we were coming."

Schroeder notes the overwhelming genetic evidence for common ancestry, yet points out a problem evolutionists face in explaining the branches of the tree. If there is no guiding hand, or purpose, or goal, in the creation of organisms, how does one explain the separate genetic evolution that produced such things as eyes? If natural selection works on mutations to maximize the organism's survival, what happens when mutations offer no advantage and only down the line are expressed? "The gene that controls the development of eyes was programmed into life at the level below the Cambrian," writes Schroeder. "That level is either the amorphous sponge-like Ediacarans or one-celled protozoa. But neither has eyes."

You may see this is not unlike Michael Behe's problem with irreducible complexity. If the precursors that lead to the bacterial flagellum have no function until part of the motor, why would they have been selected for something that didn't yet exist? Or in Richard Dawkins's words, because evolution is blind, with no ability to see into the future, how could it anticipate a future design, which, until created, has no purpose?

Convergent evolution refers to the emergence of observably similar features in different animals that, by the fossil record, we know evolved separately. Thus the eye of human beings and the eye of mollusks—having diverged half a billion years ago—represent convergent evolution in that each developed along separate branches of the tree. So the question is: how is it that random process, by which evolution is said to happen, should coincidentally produce similar outcomes without some prior plan or design?

Natural selection may indeed be systematic in its work on random mutations, and therefore can account for some apparent design as an autonomous process. Yet there remain these questions to which ID advocates await scientific answers. In fact, the conflict between Intelligent Design and evolution may move beyond the realm of science: expressing the deep-seeded human intuition that the universe must have been designed.

Evolution as Science

At a God & Science meeting at the Episcopal Center at the University of Massachusetts, a guest speaker and evolutionary biologist was explaining to me the concept of survival of the fittest. I asked him whether he thought jumping back onto a curb to avoid being hit by a bus was an example of the human evolutionary instinct to realize its own survival. He answered, of course, it was—and that even before one is conscious of such physical reaction, the instinct has already taken over to preserve the organism.

I went on to ask him what he would say if I had sacrificed my life to save my 4-year-old who had run into the street to be imperiled by the same careening bus? He said, that too is evolution's instinct to survive, insofar as it was one of my children—who represents my own genetic survival (and mine, once I am dead and gone). In Dawkins's terms, my son is the host of my genes and so I'm only acting selfishly: survival of the fittest acts at the genetic level, beneath my own living organism.

The trouble is, both answers vindicate evolution. Whatever one might think about inevitable self-interest—even in acts of self-sacrifice—however contradictory these two scenarios are, they both finally lead to evolution. This represents a problem to evolution's power to describe biological life: if all evidence proves the case for evolution, then the theory only proves itself.

Evolution Happens

That said, many ID advocates believe that evolution happens. Here it is a matter of degree more than a matter of challenging the fact that it happens. The peppered moth, of which we heard several chapters ago, is an example of microevolution—with which Intelligent Design generally has no problem, as proponent Phillip Johnson explains.

Fact or Faith _____

Nobody doubts that evolution occurs, in the narrow sense that certain changes happen naturally ... Examples of this kind allow Darwinists to assert as beyond question that 'evolution is a fact,' and that natural selection is an important directing force in evolution. If they mean only that evolution of a sort has been known to occur, and that natural selection has observable effects upon the distribution of characteristics in a population, then there really is nothing to dispute.

—Phillip Johnson, *Evolution as Dogma: The Establishment of Naturalism*

"The important claim of 'evolution,' however, is not that limited changes occur in populations due to differences in survival rates," Johnson writes. "It is that we can extrapolate from the very modest amount of evolution that can actually be observed to a grand theory that explains how moths, trees, and scientific observers came to exist in the first place."

Thus for some, evolution appears to occur only within a species. For others, such as Michael Behe—who says he has no reason to doubt common descent—macroevolution is irrelevant to the discussion of intelligent design. The point is, like any large movement composed of a variety of points of view, ID represents a diversity of opinions on the subject.

Evolution Is Not Truly Scientific

In the heat of this highly politicized battle, it may be no surprise that ID proponents would lob back the grenade that evolution isn't scientific. We have considered—and we will look at further—the question of what constitutes science; but for now we will field the ID complaint that evolution fails the scientific standard. If indeed Darwin redefined science in exclusively materialistic terms, it follows that his theory should adhere to this same standard (by which it criticizes its opponents).

Yet by the very nature of evolution, the theory is not repeatable. As a description of the past, it's a retrospective explanation that cannot be proven; and so in the same way, by virtue of its nonpredictive power, it cannot be falsified. Predictive theories must be open to testing by the chance of obtaining a negative outcome—but evolution looks back to explain what may have happened rather than ahead to what might happen.

For ID advocates, the fossil record doesn't hold evolution to account. When an unpredicted fossil appears, the theory itself is never challenged—rather, just-so stories are carefully evolved so as not to disturb the foundation. And because random mutation and natural selection cannot be observed in a lab, the evolutionary mechanisms that are claimed have yet to be adequately proven.

A Problem of Culture?

In the next chapter, we turn to the cultural setting of the ID debate. Many parallels are drawn between the Scopes monkey trial of 1925 and the recent school board trial dealing with Intelligent Design in Dover, Pennsylvania.

It is apparent that the issue has not only assumed religious—indeed, Christian—overtones, it is generally perceived as a political debate between liberals and conservatives.

One would be naive to assume that the evolution/Intelligent Design debate has happened in a vacuum—without political implications, or indeed, repercussions. The truth is there are those on both sides of the issue who've been hurt and whose careers have been damaged. As much as we may revere the pursuit of knowledge as a sacred enterprise, it is not above the partisan politics required to become the reigning paradigm.

What is interesting is that each side seems to believe it's on the short end of this reigning paradigm. Evolutionists lament that just when they thought Darwin had gotten his due, a reigning Christian history returned to unfairly impose its political will. In the same way, proponents of Intelligent Design perceive Darwinian evolution as a cultural force now deemed beyond reproach, under the flag of technology and science. The truth may lie somewhere in the middle—or rather, somewhere above the fears that dismiss the other side as inhuman idiots and suppress a far more interesting debate.

The Least You Need to Know

- According to neo-Darwinists, random mutation creates diversity.

- In the neo-Darwinist school, mutations can result in organisms that are more fit for survival.

- To Intelligent Design proponents, the neo-Darwinist understanding of random mutation in evolution is a flawed concept.

- ID proponents contend that natural selection alone cannot explain the complexity of life.

- ID advocates contend that evolution may not itself be science.

Chapter 18

A Question of Culture

In This Chapter

- ◆ The perceptions of Anglo-Americans
- ◆ The views of ID
- ◆ The views of the Darwinists

In focusing on the science—or as some may say, the pseudo-science—of the Intelligent Design debate, we could be led to believe that the conflict engendered by ID is strictly academic. Of course, it is not; just as those involved in the debate are not strictly academic. In fact we have seen how the majority of us perceive design in the universe, and how the culture has been moved to explore this question in unprecedented ways. However one chooses to answer the question of Intelligent Design, the fact is it grips us like few questions of this nature—scientific or philosophical.

Though we often hide behind constructed arguments to define ourselves in the world, such arguments do not begin to embody the complexity of who we are. Because we are no less than a vast and complex matrix of encounters and ideas, we're not always aware of from whence ideas come or how they resonate in other people. So it is important to recognize the critical importance of the ID debate—as a means by which to clarify what each of us believes within ourselves and in relationship to others.

For as much as Christianity is often blamed for demanding blind obedience, it should be argued that the term *blind obedience* is an oxymoron, self-contradictory. The word *obedience* comes from the Greek root *obeir*, which literally means "to listen"—which I have learned seldom happens in the church, in universities, and in the larger social context.

Fact or Faith _____

Satirist Fran Liebowitz observes that the opposite of talking isn't listening. "In twentieth-century American culture, the opposite of talking is waiting."

Perhaps the reason the debate between mainstream science and Intelligent Design is so heated is that it challenges not only what we think we know, but finally who we think we are. For some ID proponents, this is no less than a matter of finding their identity in God; yet for evolutionists, for whom God may not exist, the truth is every bit as important. As products of a history comprised of Christianity, the American Revolution, and Darwin, we've been given an unusual opportunity to wrestle with its fascinating questions.

Anglo-American Views

It's often said that Intelligent Design is a uniquely American phenomenon. The logic is that the United States is religiously naïve, and that Europe (specifically Britain)

Fact or Faith _____

Twenty-two percent of the British surveyed about the question of evolution opted for creationism as the preferred explanation of the origins of life. Seventeen percent chose Intelligent Design to explain our organic beginnings, whereas 48 percent attributed life to the theory of evolution.

is more sophisticated in its acceptance of evolution. However, in 2006, a poll was taken by the British Broadcasting Corporation (BBC), which found that although the British are less theistic in their science, the disparity is not what had been thought. About half of Britons surveyed recognized evolution as the best explanation, whereas almost 40 percent saw God as a part of the human creation equation.

In a 2005 poll conducted by CBS, 51 percent of Americans surveyed said they believe that God created human beings in their present physical form. Thirty percent said they believe humans evolved but God guided the process. Only 15 percent attributed life to evolution alone, leaving 81 percent recognizing God as the universe's source and creator.

Surprising as may have been the results of the BBC poll concerning evolution, the difference between the Britons and Americans still appears to be significant. Because we are affected by the culture into which we are born, develop, and mature, in the case of such beliefs, we shouldn't be surprised that they would manifest this disparity. One might identify several factors that are unique to the United States that contribute to the greater religious commitment evident among Americans.

A Country Founded on Religious Freedom

A friend, particle physicist and Anglican priest Sir John Polkinghorne, once explained to me how his priestly role in British society was different from mine. Unlike in America, where parish denotes a particular church to which one belongs, in Britain a parish is a geographical area—which by its nature includes everyone. One doesn't choose a parish, but a parish chooses all who live in a particular place; so like it or not, every Briton necessarily belongs to a parish.

Thinking of this as a quaint technicality, I asked him what difference it made. He said that when he heard someone had died—including those who'd never darkened the church door—he visited the family as members of his parish, without regard for their chosen membership. Compare this to the American cultural ethos, where church membership is a choice; indeed, such an intrusion by an unknown minister would likely be seen as a violation.

The point is that the United States was founded in the human pursuit of religious freedom—and distant as this seems, it is a cultural feature different from its motherland. The revolution went both ways: ensuring both the liberty to believe or not believe in God; one could argue that this history requires of us a choice that a more cultural religion does not. In the ID debate, Americans are perpetually reminded that God is a free choice—and an important investment in defining who they are as democratic citizens.

A Company of Deists

Deism played a critical role in the founding of this country. Espoused by Thomas Jefferson, Benjamin Franklin, and popularized by Thomas Paine, deism holds that reason rather than revelation or tradition, should motivate religious belief. In a sense it contradicts both extremes of the Intelligent Design debate: rejecting organized religion while still recognizing a transcendent deity.

In fact, most proponents of Intelligent Design resist deistic thought. It may not be so much deism's resistance to the institutional church as that ID proponents are inclined to find its God too distant from human affairs. Indeed, for some ID advocates, deism amounts to virtual atheism—insofar as God isn't seen to intervene in the world in any ongoing way.

def•i•ni•tion

Deism is "a movement or system of thought advocating natural religion, emphasizing morality, and in the 18th century denying the interference of the Creator with the laws of the universe."—From Merriam-Webster's Online Dictionary, 2006

The irony in the ID resistance to this American deistic vision is that the watchmaker argument, put forth by William Paley, is classic deistic thought. Indeed, it is this argument to which Briton Richard Dawkins devotes a 300-page book in an effort to dispel the deistic notion that the universe was designed by God. Thus deism seems to get it from both sides—from evolutionists, and ID proponents—testifying to the peculiar nature of this historical American conflict.

A Matter of Choice

In the context of American history however, the debate makes utter sense. If the content of the conflict has its roots in the American quest for religious freedom, its rhetorical form has its roots in the free speech guaranteed by our democracy. Rather than dismissing the nature of the argument through the eyes of other cultures, we might see the controversy about Intelligent Design through the lens of respect for ourselves.

This is not to say the substantive scientific issues are irrelevant. In fact, it is to say they're of supreme relevance—in this human pursuit of the truth. Yet as we shall see, the conflict is often not a matter of pursuing truth, but an opportunity to ventilate resentment, or to decimate the opponent.

The question of design is more interesting an issue than most people have been willing to allow. Academics, politicians, the courts, and the media generally assume it's a nonissue: mere religion, a masked conspiracy, that should be instinctively dismissed. Some of these judgments may indeed be right—that ID is used for deceitful ends. Yet is this therefore to say that more than two millennia of philosophy should also be dismissed?

The ID View

Proponents of Intelligent Design believe they've been unfairly excluded by the academy, the public schools, the media, the courts, and indeed, by the culture at large. We shall see how their opponents equally believe their own positions are being threatened, and that they must do all they can to counteract the falsehoods being perpetrated by the movement. But for now it is important to understand the common perception held by ID proponents is that rampant secularism has effectively expunged religious belief from the culture.

Phillip Johnson, a retired law professor from the University of California at Berkeley, is often recognized as the spiritual father of the ID movement itself. Johnson is seen as the political strategist inside and outside of the movement, having forged the now famous Wedge Document, which lays out ID's political approach.

> ### Evolutionary Revelations
>
> The two governing goals of the Wedge Document are (1) to defeat scientific materialism and its destructive moral, cultural, and political legacies; (2) to replace materialistic explanations with the theistic understanding that nature and human beings are created by God.

The Wedge Document has fueled suspicions that ID is more than scientific inference: some perceive it as a stealth conspiracy to bring religion into the public schools.

The recent court decision *Tammy Kitzmiller v Dover (Pennsylvania) Area School District* ruled in favor of keeping Intelligent Design out of the curriculum. For mainstream evolutionists, this was seen as a victory over ID advocates who sought to compromise scientific education and the separation of church and state. To proponents of Intelligent Design, Judge John E. Jones's decision was one more expression of how secular culture denies any transcendent explanation.

Religious Atheism

Phillip Johnson has argued that although science stands as its own legitimate discipline, when it becomes a philosophical worldview, it oversteps its bounds. Mainstream scientists respond that all they are seeking is respect for their methodology, letting philosophy and religion have their due—just not in the laboratory. The problem, according to the ID perspective, is these two disciplines are not so distinct—specifically, what passes for science is often an atheistic point of view.

The point is that all knowledge has philosophical implications. Despite potential claims of scientific objectivity, Darwin's materialism can lead one to focus on natural explanations that may imply a Godless universe. Though we are about to find out that there are those who are able to embrace both God and evolution, for some, to deny there is design in the universe is to deny divine creation.

> **Blinded By Science** _____
>
> The problem with scientific naturalism as a worldview is that it takes a sound methodological premise of natural science and transforms it into a dogmatic statement about the nature of the universe. Science is committed by definition to empiricism, by which I mean that scientists seek to find truth by observation, experiment, and calculation rather than by studying sacred books or achieving mystical states of mind. It may well be, however, that there are certain questions—important questions, ones to which we desperately want to know answers—that cannot be answered by the methods available to science. These may include not only broad philosophical issues such as whether the universe has a purpose, but also questions we have become accustomed to think of as empirical, such as how life first began or how complex biological systems were put together.
>
> —Phillip Johnson, "Evolution as Dogma," _First Things magazine,_ 1990

The Courts, the Media, and the Academy

In the Kitzmiller case, Judge Jones asserted that Intelligent Design was in fact a version of creationism that rejects evolutionary theory. What is interesting about Jones's assertion is that principal witness, Michael Behe (of irreducible complexity fame), says he believes in evolution. As we have seen, creationism holds to the biblical creation story, whereas evolution requires that plants and animals were not created in their present form.

The media also seems unwilling to take ID arguments at face value; instead, its coverage of Intelligent Design invariably presumes creationism. So deep are the suspicions that Intelligent Design means God (which for many, it does) that the ID arguments are rarely described beyond religious motivation. In the same way that a godless universe would have been rejected in Galileo's time, now this ancient theistic tradition has been rejected as unscientific.

Academic tenure was originally conceived in part to grant faculty protection—indeed, it proved its political usefulness during the McCarthy era. In the sciences, however, tenure may have seemed like an unnecessary artifact: at least, until Intelligent Design shed its light upon the politics of laboratories. Many of the leading ID proponents have lost their positions and their grants; and however scientifically competent they are, they have little hope of being published.

Of course cases can be complicated and deserve to be judged on their own particular merits. However, one would be naïve to assume that science escapes vicious politics.

Just as religious and political institutions have always been about power, so are corporations and academic institutions—sometimes at the cost of larger truth.

The Wedge

In response to the unfairness ID feels it suffers at the hands of mainstream scientists, Intelligent Design has clearly become a highly politicized movement. The Discovery Institute is given credit for authoring the Wedge Document—an initiative "to replace materialism and its destructive cultural legacies with a positive scientific alternative." (See appendix, *The Wedge Project*.) Though some contend the Wedge Document was made public by a disgruntled institute employee, the institute is unapologetic about its mission to American culture.

Evolutionary Revelations

The social consequences of materialism have been devastating. As symptoms, those consequences are certainly worth treating. However, we are convinced that in order to defeat materialism, we must cut it off at its source. That source is scientific materialism. This is precisely our strategy. If we view the predominant materialistic science as a giant tree, our strategy is intended to function as a "wedge" that, while relatively small, can split the trunk when applied at its weakest points.

—*The Wedge Project*, Discovery Institute

One might see as much ideology in this apparently anti-science statement as in the most strident atheistic mainstream scientific claims. If Intelligent Design is genuine in its desire to make minimalist claims—that is, claims in favor of inferred design rather than against materialism—it would seem that its politicization as a movement could compromise its pursuit of the truth, miring it in the selfsame power hungry interests to which it believes it has fallen prey. Indeed neither side appears to be wholly innocent in this regard; in both cases, it would seem that ideology has stood in the way of good science.

The Darwinian View

Mainstream science is not immune to affecting a surrounding culture or developing its own doctrinal beliefs that transcend its work in the lab. Although mainstream science is not a singular movement (neither is Intelligent Design), it can be given to evolve a worldview to which individuals naturally assent. What may be almost universally agreed by mainstream scientists is that appeals to the transcendent, or the supernatural, cannot be a part of the process.

In this scientific day and age, such an assertion seems almost self-evident—a tautology, which proves itself by virtue of the manifest nature of science. As we have seen, however, this wasn't always the case, but was largely occasioned by Darwin; it was only in the middle of the nineteenth century that godless science came to the fore. One could argue, in as much as Galileo and Newton inferred God in the universe, so modern mainstream science uses its own knowledge to infer that God does not exist.

Atheistic Religion

Perhaps the most obvious examples of ideology exceeding the parameters of science are zoologist Richard Dawkins and philosopher Daniel Dennett. We have discussed their theologies of atheism for their philosophical content; but it is also important to recognize their roles in the culture of the ID debate.

In the same way the Wedge Document incites anger among mainstream scientists, so have the remarks of Dawkins and Dennett among proponents of Intelligent Design. We must recognize when scientific discussion is no longer scientific and when it becomes a philosophical debate about the existence of God. This is not to say such debate is wrong, but only that it's not about science—and that if debate is for the sake of greater truth, it shouldn't be about an enemy. In Richard Dawkins's case, the enemy is clear.

Insofar as for Dawkins all of life is merely a function of evolution, one has to wonder why, blind as this life is, he should begrudge the existence of disease.

Daniel Dennett (who has said that religious people belong in cultural zoos) has just completed a book in which he argues that religion is an evolutionary adaptation. In the same way that Dawkins speaks about religion in seemingly scientific terms, Dennett similarly uses scientism to argue his case for atheism. In fact, some orthodox Darwinians criticize contempt by scientists for religion—not so much on ethical grounds as on the philosophical ground of genuine science.

> **Evolutionary Revelations**
>
> It is fashionable to wax apocalyptic about the threat to humanity posed by the AIDS virus, 'mad cow' disease, and many others, but I think a case can be made that faith is one of the world's great evils, comparable to the small pox virus but harder to eradicate.
>
> —Richard Dawkins, "Is Science a Religion?" *The Humanist*, January-February 1997

Politics, Not Science

Not all critics of Intelligent Design have a theological bone to pick. Barbara Forrest, a philosophy professor at Southeastern Louisiana University, explores possibilities for human meaning in the theory of evolution. However, she views Intelligent Design as a religious and political movement rather than a genuine school of thought in pursuit of scientific truth.

Referring to the Wedge Document, Forrest quotes one of Phillip Johnson's philosophical confessions, and another leading voice, William Dembski, about his Christian presuppositions.

"The Wedge aims to 'renew' American culture by grounding society's major institutions, especially education, in evangelical religion," Forrest writes in the April 2002 edition of Natural History magazine. "In 1996, Johnson declared: 'This isn't really, and never has been, a debate about science. It's about religion and philosophy.' According to Dembski, intelligent design 'is just the Logos of John's Gospel restated in the idiom of information theory.' Wedge strategists seek to unify Christians through a shared belief in 'mere' creation, aiming—in Dembski's words—'at defeating naturalism and its consequences.' This enables intelligent-design proponents to coexist in a big tent with other creationists who explicitly base their beliefs on a literal interpretation of Genesis."

Forrest expresses Intelligent Design as a monolithic movement whose actors consort to further a uniform political-religious agenda. She equates Dembski's call "to devote [his] life to destroying Darwinism" with Michael Behe's quite different philosophical question: whether science can make room for religion. Although the first critique highlights the political character of some of the ID movement, the question Behe poses may be a more historically relevant scientific question.

Necessary Respect

In an e-mail sent to Dennett (and since published at www.uncommondescent.com), philosopher of science Michael Ruse writes, "I think that you and Richard (Dawkins) are absolute disasters in the fight against intelligent design—we are losing this battle, not the least of which is the two supreme court justices who are certainly going to vote to let it into classrooms—what we need is not knee-jerk atheism but serious grappling with the issues—neither of you are willing to study Christianity seriously and to engage with the ideas—it is just plain silly and grotesquely immoral to claim that

Christianity is simply a force for evil, as Richard claims—more than this, we are in a fight, and we need to make allies in the fight, not simply alienate everyone of good will."

Michael Ruse is no friend of Intelligent Design and identifies himself as an agnostic. He is in fact an ardent Darwinian evolutionist—of the orthodox stripe (rather than of later neo-Darwinist sociologies).

Of all the thoroughgoing evolutionists, Ruse has modeled scholastic engagement: engaging scientific issues and co-editing (with an ID proponent) *Debating Design*. To the extent the ID conflict remains a self-centered, political struggle for power, the substantive science and philosophy will inevitably suffer. If it remains a dogmatic restating of predetermined philosophical positions, the knowledge to be gained by the other side will be lost to what we already knew.

The Least You Need to Know

- Based on polls by major media companies, Britons and Americans differ in their perceptions about evolution, creationism, and Intelligent Design, although perhaps not as much as might have been imagined.

- America's perceptions on both sides of the debate are undoubtedly influenced by America having been founded on the freedoms of speech and religious expression.

- Many Intelligent Design proponents believe that ID gets short shrift in the debate as a result of our modern culture's increasing secularism.

- In the case of *Tammy Kitzmiller v. Dover (Pennsylvania) Area School District*, the judge ruled that ID was simply a version of creationism.

- The Wedge Project is a Discovery Institute initiative that the institute claims is aimed at replacing materialism with a scientific alternative.

- The heated Intelligent Design debate involves questions concerning science, culture, and politics.

Part 7

Is a Resolution Possible?

What is the future of the debate over Intelligent Design? It is obviously contentious and fraught with obstacles at the present moment.

Here we look at possible resolutions and potential common ground. We consider different approaches to science to better understand and appreciate the possibility of Intelligent Design.

Chapter 19

The Nature of Knowledge

In This Chapter

- What is meant by "understanding"
- The arguments by "science"
- What exactly is science?
- Who or what is the authority?

There can be no real constructive engagement in the Intelligent Design debate unless there is some common agreement about the nature of science itself. In the absence of such common ground, or the apparent desire to gain it, the debate, as a way to truth and knowledge, is certainly destined to fail. In short, as long as "the opposite of talking is waiting" (as opposed to "listening"), there will be no resolution in the Intelligent Design debate.

The Nature of Understanding

Socrates taught that human knowledge comes by humility—rather than pride, which has come to characterize the present discourse. In this regard, it's not just a matter of getting the science "right": it is also a matter of understanding the nature of human knowledge.

Rather than emptily claiming that something is "a scientific fact," we must say what we mean by "scientific"—and what we mean by a "fact."

What Is Fact?

When we say something is a fact, we normally mean that whatever "it" is, is true. When we say something is scientific, we normally mean whatever "it" is, is proven. Yet this says little, except that what we think is true, is true—leaving both sides without any traction—by which to explore the substance of the conflict, and to move the debate ahead.

If Intelligent Design is seen to lie beyond the pale of science, the question of what science is (and is not) must be actively pursued. If mainstream science as currently conducted is necessarily agnostic—that is, requires that "God" be excluded from the scientific equation—this should be decided with an eye to the thousands of years of theistic science and explained for its inherent "improvement" over its God-fearing predecessors. Arguments can be made for both sides—for theistic and agnostic science—but are rarely made or heard in the clamor of the ID battle.

What Is Truth?

What is science? How does it differ from theological study? What do science and theology share as means of pursuing truth? These are fascinating questions that have not been fully answered—to the detriment of both disciplines.

Fact or Faith _____

In his poem "Ash Wednesday," T. S. Eliot says, "Teach us to care, and not to care ...." Although Eliot is apparently speaking to God, one could imagine him addressing a world that was beckoning those who fearlessly cared to finally know the truth. For if the truth is the truth, we needn't worry, or somehow try to change it—it is simply there for our understanding, whether we receive it or not.

There is a story about Niels Bohr hosting several other celebrated physicists at his summer cabin in Norway. As his friends were arriving, one of them noticed a horseshoe nailed to the lintel above the front door and teased his host, "Come, Niels, surely you don't believe in that!" To which Bohr allegedly replied, "What does that matter? It works, whether I believe in it or not."

What is instructive here isn't whether horseshoes really bring luck. What is instructive is that whatever our beliefs, if they are true, we don't have to protect them. And if they are false, they become the means by which to journey toward greater truth—leaving us with nothing to lose, except perhaps our ignorance.

Of course this is easier said than done. It requires enormous discipline, conviction, and trust in the truth—a conviction that only comes by way of risking oneself to believe. Yet there are a few in the ID debate who model this kind of conviction which is required to move forward the conversation past the curtain of ignorance.

The Scientific Arguments

As frustrated as mainstream scientists seem (particularly evolutionists) in answering criticisms launched by proponents of Intelligent Design, it appears that however satisfied they are with their own answers, more than half the public still needs to have its dissatisfaction addressed. As long as mainstream science contends ID proponents are just "creationists"—who refuse to listen to "the evidence, which needs no further explanation"—the majority of America will continue to live in "ignorance," and "the ignorant" will continue to feel disregarded and dismissed.

As educators, mainstream scientists must engage in effective education; their continuing frustration testifies to the truth of their own failure.

In the same way the debate requires mainstream scientists to listen to ID proponents, these proponents must respect mainstream arguments to approach a greater truth. If either side assumes that it will emerge unchanged by the other, such insincerity will doom the discourse from the start. Whether speaking about the substantive issues, the nature of the science, or surrounding politics, only a climate of openness will render new understanding.

Argument from Ignorance

We have seen the major arguments put forth by leading ID voices as a way to theorize the universe, in its complexity, was designed. And we have seen some mainstream scientific answers, which equally seek to explain the complexity of the universe in exclusively material terms. To the extent that complexity seems to be a feature about which there is agreement, it might be a fruitful starting point for the greater conversation.

def•i•ni•tion

The word **complexity** is a problematic word in the current ID debate. In the media, one often sees Intelligent Design roughly defined as follows: the theory that life is so complex that it can only be explained by an intelligent designer. From the perspective of Intelligent Design, this is a misuse of the term *complexity*.

The word *complexity* is used by Michael Behe (and others) in a technical sense. Remember that Behe described irreducible complexity as "a single system composed of several well-matched, interacting parts that contribute to the basic function, wherein the removal of any one of the parts causes the system to effectively cease functioning." Here, complexity is a technical description of functional interdependence.

This is different from what many describe as an "argument from ignorance." An argument from ignorance essentially asserts that something is so complicated that we can't possibly explain it, and that therefore God must be at the bottom of it. There are undoubtedly those who believe this is so and have no interest in further exploration; however, this is not Behe's position, whose design inference comes from "the stuff."

Argument from Knowledge

Arguments from ignorance doom conversations between Intelligent Design and mainstream science. Because both sides appeal to the material stuff to make their arguments, all arguments must be embedded in their empirical research. What is ironic is that each side blames the other for conducting inadequate science—leaving lay observers to draw their own conclusions, often "unscientifically."

From the mainstream perspective, Intelligent Design disregards scientific knowledge. So biologist Ken Miller critiques the idea of "irreducible complexity," invoking "precursors," which he believes lead to the evolution of the flagellum. Indeed, Miller believes that Intelligent Design is blind to such explanations due to its nonscientific, and indeed, theological presuppositions.

In this regard, mainstream science contends that the problem with Intelligent Design is that, given its inherent theological nature, it can't have a research agenda. If the inference is "God," or some "transcendent agent," there is no way to test its existence, and therefore there is nothing about which to agree—or indeed, to disagree. Here we are reminded of Karl Popper's definition of science by falsification: if the hallmark of science is "falsifiability," how does one falsify God?

To which Behe responds that scientists try to falsify his arguments all the time. The bacterial flagellum as irreducibly complex gave rise to Kenneth Miller's objection that precursors and other evidence indicate the flagellum was evolved. We have covered the content of this argument—what is important to recognize here is that there seems to be room for more substantive debate than has been allowed until now.

Evolutionary Revelations

When I asked Michael Behe about the credibility of Miller's explanations, he aired his frequent frustration that most of them amounted to "hand waving." The point is the presence of disagreement about material processes indicates the need for a scientific discourse that fosters accountability. However short of naming "the transcendent agent," or identifying the "God," the stuff from which Intelligent Design draws its conclusions can be scientifically debated.

Though there may be ID advocates who strategically hide their theological convictions, there are those who recognize there is nothing in the "stuff" to indicate "who" designed it. Behe insists (with regard to the specific identity of a designer) that Intelligent Design necessarily makes only "minimalist" claims. Yet what might be determined is whether or not an organism required design—deducing no more than that the agent of design exhibits "intelligence."

To Overlap or Not

One of the questions that began this *Complete Idiot's Guide* was the question of theology and science: whether these two very different disciplines should have any interaction with each other. By its very nature, Intelligent Design necessarily asserts that they must, whereas most mainstream scientists, by present definition, assert they must remain distinct. We might recall the argument of Stephen Jay Gould about the relationship between God and science: that by virtue of their respective differences, they're "nonoverlapping magisteria."

The question is, why is this not satisfying to the general population—including even that of Great Britain, where Charles Darwin was born? It would seem that most people outside the lab perceive some relationship, and are even inclined to propose a creator is responsible for the universe. Although proponents of ID would explain this is because the universe *was* designed, many Darwinists fault bad education or the complexities of science.

Advocates of Intelligent Design often contend that this separation is symptomatic of the way evolution has emerged as the reigning religion. You may remember it was Darwin who erased God from the scientific perspective, showing how material processes could operate on their own. So one of the criticisms of Intelligent Design, which is made by mainstream science, has been that Intelligent Design is a religious movement—or at least, that ID isn't "science."

Yet as we have seen, what passes as science has dramatically changed over time—and unless our time proves historically unique, it will continue to change. What is interesting is how little this change has been publicly recognized, giving us to think the word *science* has always borne an absolute, unchanging meaning. The truth is the word *science* is frequently used less for its methodology than for the exercise of power—in the academy, the government, and the culture at large.

Definitions of Science

Judge Jones's judgment in Dover, Pennsylvania, was based on the prior assumption that the word *science* could be clearly defined, and so distinguished from theological assertions. But as we have seen, science and theology were wedded for centuries, and only with Darwin did science become an agnostic enterprise. This is not to say that therefore science must return to its theistic roots; it is only to say that Intelligent Design is no historical quirk.

At a recent meeting of the God and Science Project, biologist, Steven Goodwin was introducing Michael Behe (to whom he would respond from his mainstream scientific perspective). At the end of his remarks, casting doubt on ID, Goodwin honestly confessed that for all the authoritative claims in the air, there's no good definition of science. Which returns us to the problem faced by the American Association for the Advancement of Science, which struggled to develop a working definition and finally had to give up.

If the Intelligent Design debate is ever to move in fruitful directions, it will have to wrestle with questions related to the philosophy of science. It is clear the dispute is not only about the politics of power; on both sides of the issue is genuine concern for the content of their beliefs. Yet the nature of knowledge, of material reality, and of the possibility of God, all affect the way we perceive the role of science in pursuit of the truth.

Materialism

From a general perspective of Intelligent Design, it is easy to make Darwin a villain. But the truth is that Darwin was no scheming atheist out to change science forever. If he was a complex combination of atheism and religious faith, it was the credibility of evolution that squeezed God from the scientific method.

Keeping God out of the method has become a goal of modern mainstream science. As has been said, to conclude there is no God is a matter of philosophy—however it might parade as a thoroughly scientific deduction.

Excluding God is significant. More than implying the material world is "the only game in town," it also means God can't be inferred from the surrounding physical world. So the world is no longer a way for believers to apprehend their "God"—leaving them to question God's credibility as the source of material creation.

Fact or Faith _____

The materialism presumed by mainstream science need not preclude a God outside it; it is only that it can't be seen as an active part of the empirical process.

One might argue that this materialist perspective is as great a threat to art as it is to religious experience, or to the possibility of God. To deny the transcendent in the material world—in paintings, in books, in music—is to deny there is significance beyond the physical stuff of which it is made. Because art depends on generating meaning which transcends its physical incarnation, a critical, mutual dependence exists between the physical and spiritual realms.

Design

The scientific revolution begun by Copernicus was fueled by philosophy—by neo-Platonists who worshiped the sun as the center of the universe. Galileo would wax poetic about the universe's design, never seeing conflict between the God and the heavens he deeply revered. And although Newton was known as the mechanist, whose world was ruled by autonomous laws, he believed that God tinkered with the planetary orbits to make his calculations come out right.

Even after Darwin, the transcendent was part of the scientific equation. Dethroning the Newtonian paradigm, Einstein saw new majesty in the manifest design he seemed to attribute to some kind of deistic designer. As much as relativity theory might have upset Newton's mechanisms, Einstein's remark that God didn't play dice still asserted God as a mechanist.

And we have seen how the big bang—filled as it is with theological implications—resolved the ancient question of how there could be the uncaused "first cause" of Aquinas. The theory presumes that all physical laws—including causality—didn't come into play until the seminal moment of creation. So there is no need for an uncaused first cause (which itself is a contradiction), because only when God's world exploded into being did such laws have to apply.

Evolutionary Revelations

So one might imagine an infinite realm in which God alone would dwell—without the constrictions of a finite world that is bound by physical laws. Such quandaries as "divine omnipotence" could be resolved in scientific terms: if God is all powerful, how could he lift the greatest weight he could possibly create? By way of the big bang, theology and science might find a resonance that opens up new understandings of God and the physical universe.

Uncertainty

What do evolution and Intelligent Design philosophically have in common? Something they seem to share is the view that the world is mechanistic—whether natural selection, or an intelligent agent, mechanisms make for life. If we're uncertain, it is because we haven't figured out the mechanisms yet—but given enough time and research, one day we will isolate the dynamic.

But as we saw in the chapter on mainstream physics, the theory of quantum mechanics says there are features of the universe that are necessarily uncertain. Nothing we do can ever change this condition—not technology, nor better knowledge—the fact is that, at the particle level, lies an uncertain universe. So the certainty presumed by mechanistic views of God and evolution must be reconsidered in this "uncertain" light, with all its theological implications.

What do we do with our conceptions of God which depend upon certainty? Although uncertainty explains the chance and accidents of our daily lives, doesn't it call into question the very certainty of God? If God allows uncertainty in the universe (that therefore even God can't control), what does this say about the nature of God, and God's relevance to the universe?

These are fascinating questions which need to be answered by science and theology together, in recognition of our common pursuit of the truth that is woven into life. If fear is the greatest enemy of every pursuit of knowledge, insofar as science and religion suffer fear, they are giving themselves to defeat. But if "It's true, whether I believe it or not," then neither side has anything to lose—except, perhaps, superstitious ignorance, which keeps it from greater knowledge.

Who's Got the Power?

It seems clear that the culture-wide ID debate is at least about political power. In government, Intelligent Design is contested in the White House and on Capitol Hill. It's being argued in the courts, as a result of being introduced into the schools—with the ACLU claiming that it threatens the separation of church and state.

But the politics of power are not only a matter of governmental institutions. They are also a matter of scientific institutions which award grants and publications. They are a matter of the classroom, and the laboratory, by which academic tenure is awarded—and where opinions are made, where minds are shaped, and where value-laden worldviews are conceived.

Fact or Faith

The American Civil Liberties Union today applauded a court ruling that it is unconstitutional to teach "intelligent design" as an alternative to evolution in a public school science classroom.

—American Civil Liberties Union, press release, December 20, 2005

They are a matter of the nature of the media, by which it will suffer or thrive. They are a matter of the church, which feels its power slipping from the days when it was to be reckoned with—in many cases, facing its social demise at the hands of this vastly changing time. Therefore, in the culture—in its myriad expressions of fear, uncertainty, and conflict—Intelligent Design is acting as a flash point for our most deeply held beliefs.

The Political Arena

One need only open the newspaper to find the latest on Intelligent Design. As Michael Ruse asserts, it may have implications for upcoming Supreme Court appointments. Legal precedents are being established throughout the judicial system—even reshaping the role of religion in the public context.

Fact or Faith

I once asked a leading critic of Intelligent Design what it was he didn't like about ID—thinking he would talk (as a philosopher) about the nature of scientific knowledge. Instead, he said he didn't like its "politics" (presuming them to be conservative), giving me to realize how affected we can be by what seem to be peripheral issues. It is important to divorce—as much as we can—the politics surrounding ID in order to assess the arguments on their own scientific merit.

How we negotiate this contentious issue will determine how we navigate our lives in this conflict-ridden time, which seems, at every turn, to be questioning what we thought we knew. At the bottom seems to be the question of "truth"—in politics, in science, in religion—positioning ID to catalyze (for many) their most fundamental beliefs.

Renowned theologian, Paul Tillich, defined religious faith in terms of "ultimate concern"; so it may be ours to utilize ID as a lens through which to see our deepest faith.

The Academy

The stakes are high in academic politics. In fact, many scientists dependent upon grants have been ruined by their ID positions. Tenure, dependent upon mainstream publications, can be withheld by a jury of their peers. As with any institution, politics pervade colleges and universities—requiring, "academic," if not "political correctness"—in departments, in the classrooms, in the halls.

The stakes of an academic debate may also be high in the culture. The evolution/ID controversy is one such example of high stakes. The culture has weighed in—the government, the courts, the schools, and the media—therefore many eyes have turned to the academy to help to clarify the issue.

The problem is cool heads don't always prevail, and in the e-mail message seen in the last chapter, Daniel Dennett engaged Michael Ruse by threat of no less than academic exile. Because of Ruse's own tenured position, he appeared to be unfazed; nevertheless, the politics were clear, and could jeopardize those of lesser standing. Ruse responded to Dennett's e-mail by first back quoting Dennett: "You may want to extricate yourself, since you are certainly losing ground fast in the evolutionary community that I am in touch with."

Ruse then responded, "I am a full professor with tenure at a university known chiefly for its prowess on the football field, living out my retirement years in the sunshine—I have no reputation to preserve, and frankly can say and do whatever the f--- I want to without sinking further."

But Michael Behe no longer receives grants and cannot maintain a lab. ID proponent William Dembski, a Cambridge-trained mathematician, was removed as the head of the Michael Polanyi Center at Baylor for such politics. Numerous other academics have paid the cost of Intelligent Design—having lost their positions or their academic standing in their respective departments.

The Church

Oddly, the political power currently surrounding the ID debate appears far less centralized in the church than in our academic institutions. Although the Darwinian position seems relatively consistent and uniform (at least in content), the church's position spans the spectrum from creationism to mainstream science. The assumption that all Christians favor ID is far from the reality—the church seems to bear a greater diversity of opinion than the academy.

From the creationist position of Henry Morris within the evangelical tradition, to the Roman Catholic critique of Intelligent Design as only "pretending to do science," to liberal Protestants, including the Anglican archbishop of Canterbury—who implied that ID debased the Bible by subjecting it to scientific standards—the church is composed of wildly competing arguments and positions that may be more reflective of society at large than most academic institutions.

If there is agreement, it is generally that God is somehow present in the universe; the church as commanding a centralized teaching, is no more than an illusion of the past. Since the Reformation, when Protestants "protested" a singular church authority, Christians have exercised their God-given freedom to realize their particular faith.

The time in which we live seems unprecedented for its scientific exploration. For those who are courageous in pursuing their faith, while engaging their deepest doubts, scientific truths—including those of evolution—can nourish theological truth. However science challenges our once static convictions about the nature of God, those who sense God's presence in the universe must understand the universe itself.

The Least You Need to Know

- ◆ There is no single definition of science.

- ◆ Mainstream scientists and Intelligent Design proponents must first achieve common ground on a definition of science if the ID debate is to be productive.

- ◆ According to such a definition, mainstream scientists and Intelligent Design proponents must recognize and admit when they are making philosophical and/or theological assertions.

- ◆ The debate is greatly affected by perceived political, judicial, religious, and academic power.

- ◆ The topic of Intelligent Design needs to be discussed in an informed court of public opinion.

Chapter 20

Evolving Questions

- ◆ A fearful climate
- ◆ Two opposing sides
- ◆ Is evolution *always* fact?
- ◆ Dismissing religion

In the Faculty Club at the University of Massachusetts hangs a sign behind a long dining room table that proclaims "Often in error, never in doubt."

One of the features of the current ID debate, both within and outside of the academy, is that participants tend to go at it with an apparent conviction that they do not really have.

Defensive as they are in the face of other intelligent people, they dogmatically assert their positions as though it couldn't be any other way, and "if you don't think so, you're stupid." The result is the proverbial food fight we see in universities across the country, which has clearly affected the issue's treatment in the culture-at-large.

In fact, friends and colleagues at the University of Massachusetts are frequently taken aback when they hear that I am writing a book on Intelligent Design. They know that my own politically "liberal" orientation does not align with what has been identified as a politically "conservative" position.

Indeed, many of my clergy colleagues in the liberal Episcopal Church scratch their heads—as though even to be interested in the issue of ID is to have contracted leprosy.

Climate of Fear

In my interview with Michael Behe (see Appendix A), he speaks of the climate surrounding Intelligent Design as one of fear. Intelligent Design advocates are often dispensed with by the opposition, he says, not unlike the opposition was dismissed in the McCarthy era of the 1950s.

Whether or not one agrees with Behe's view, it is clear that honest engagement is always preferable to dishonest charges when striving after the truth. Having all suffered the second strategy for the last several decades of ID, perhaps it is time for us to embrace Socrates's notion (roughly put) that the wise man knows that he doesn't know, and so discipline ourselves to respectfully listen to "the other" point of view in exploring Intelligent Design.

More Than Just Explaining Complexity

I find myself saying to academic colleagues that whatever one feels about ID, it is a far more interesting question than the media and most scientists allow it to be.

For example, the common newspaper definition that "Intelligent Design is the theory that life is too complicated to be explained by evolution," needs to be more accurately portrayed if we're to ever have a productive debate. As we have seen, such is a far cry from the full breadth of Intelligent Design thinking, and necessarily distorts and dismisses the ID position to the reader.

ID Is Guilty, Too

That said, there is enormous resistance among ID advocates to open discussion.

Similarly invested in the assumption that "the opposite of talking isn't listening; it's waiting," ID proponents are equally given to making assumptions about "the other side." Tarring mainstream scientists with the brush of atheism, they steer the debate in the direction of a religious war that obscures fascinating scientific and religious questions. By mutual respect, all could be enriched by the challenge of another point of view—rather than listening to what we already believe, or indeed, "preaching to the choir."

Evolution vs. Religion

Although we have seen that the Intelligent Design position goes beyond biology, the truth is the heart of the ID debate resides in the evolution dispute. For many reasons, historical and current, evolutionary theory has been seen as the overriding threat to ID and to religious faith. Yet attempts have been made to quell the conflict by creating bridges between the two that embrace these respective points of view in one overarching perspective.

"Theistic evolution" assumes that God and evolution are compatible. Briefly put, evolution is perceived as a God-given mechanism that puts biological life into a motion that is unbroken by divine intervention.

In this light, the integrity of both God and evolution are allowed to remain in tact, and the conflict between them as competing points of view is successfully averted.

This position is not unlike the deistic view of God as a distant creator—who set the world in motion and then let it run without subsequent involvement. As satisfying as this may be to some, to others it fails to allow for God to be actively engaged in the ongoing process of creation. God may appear too vague or "unreal" in this evolutionary scheme of things, giving rise to the feeling that "he" has been effectively taken out of the picture.

Why is it that religious believers so often fear evolution? Is it simply that evolution precludes the need for a deity? Does it strike an insidious moral overtone that bothers religious believers, which they may not even be conscious of as they contemplate evolution?

All About Me

To religious people, and perhaps specifically to those in the Christian tradition, there may be something seemingly selfish in the assumptions underlying evolution. To the extent that evolution implies a godless world, humans may be seen as prideful beings who are suddenly free to do what they want, without regard for other people.

> **Evolutionary Revelations**
>
> Survival of the fittest, for many Christians, has profound moral implications. Mere survival is considered insignificant in pursuing the Christian life. Physical existence isn't enough, but is a means by which to strive for God—who himself transcends the material stuff that neo-Darwinists imply is all there is.

This may be no more true than that religious people thereby escape self-centered actions. Still the humility implied by God is, for many, a way out of narcissism.

The struggle to survive implies a competition that may be unsettling to Christians. It's true that Christianity has ignited competition: in war, business, and personal ambitions. Still there is a sense that competition finally defies Christian community. Cooperation, rather than competition, appears to be what Jesus taught—and so the separation that "survival" implies can be a threat to the Christian vision.

In fact, casualties of such competition are those who cannot compete. Among "the fittest" are those who are relatively less fit and are destined not to survive. Yet in Judeo-Christianity—indeed, one could argue, in all major religious traditions—the concern is for those on the margins of life who are least able to survive. The poor, the sick, the socially rejected are those of religious concern—and often represent the face of God and the place where God is most potently found.

In this regard, those who struggle for themselves may defy the character of God.

Selfish Gene and Selfish Joan

In his discussion of the selfish nature of genes, Richard Dawkins insists that such selfishness bears no moral implications, and is simply the way that life works. But for those who believe that life is realized in the service of others, and of God, it is difficult not to see this selfish urge in finally godless terms.

In the Hebrew scriptures and the New Testament, self-giving is the stuff of God. As a wandering people who knew what it meant to reside on the margins of the world, it was caring for each other that finally held this community of God together. And in the Gospel stories, it was always the poor and the stranger to whom Jesus went as a way to realize the vision of God, despite the Roman competition. Therefore, to say the world has been constructed to live by "survival of the fittest" stands as an assault upon the nature of God, and indeed, the very purpose of this life.

Perhaps one of the reasons (however subconscious) that such selfishness seems offensive is that for Christians, the crucifixion is an icon of God and a model for human self-giving. Rather than Jesus' life being seen as having been taken from him, it is rather to be seen as being freely given as a gift to his followers. However Christians may fail to realize this paradigm of radical self-giving, the crucifixion provides us a lens through which to see God and the way to the resurrection.

A Symbiotic Relationship?

One could easily see this conflict as a case of irreconcilable differences. Perhaps this is why "theistic evolution" remains a vague idea, which seems sloppy to hard-driving Christians no less than to hard-driving neo-Darwinists. Yet it may be precisely the starkness of the difference that affords a possibility of exploring the nature of the biological world in consort with the nature of God.

If biological life is genetically inclined to seek its own self-interest, one might naturally ask, "What might be the place in biology for self-giving?" A traditional way for Christians to perceive the theory of evolution is as a godless process whereby human beings are free to have their own self-centered way.

> **Evolutionary Revelations**
>
> As is often pointed out in the ID debate, a critical difference exists between simply describing the way that life works and prescribing the way that life could work.

But let's consider the following questions:

◆ What if "survival of the fittest"—which is so often seen as a threat to religious faith—is a necessary means by which human beings receive a transcendent God?

◆ What if the material stuff of our selves is a vehicle for our own choice: either to fulfill our private self-interest or be fulfilled by the interest of others?

◆ What if the "transcendence" attributed to God also means transcending our self-interest, allowing us to value evolution as a way to understand the God beyond ourselves?

Obviously, such questions can't be answered in one, or perhaps many conversations. Yet humbling ourselves before such open questions, in dialogue with others, may mark the profound humility that Socrates recommended. Here, "Often in error, never in doubt," can be rewritten in the Faculty Club to read: "Never in error, (because) always in doubt," in the spirit of many great thinkers.

The Fact of Evolution?

Another set of questions has to do with determining the truth of evolution. The claim that "evolution is not a theory, but a 'fact,'" is often made in the controversy. The word *fact* is often used to silence counterargument by virtue of the word's sheer weight; after fact is invoked, as in the case of evolution, there is no more reason to debate.

The problem with this idea is that fact is subject to definition and interpretation. The word is often used to say something is true beyond the shadow of a doubt. Or it is sometimes used in a more technical sense, to mean something is true from every point of view—perhaps in the spirit of Plato's "pure form," which he believed could bear the only truth.

def•i•ni•tion

According to Merriam-Webster's Online Dictionary, 2006, a **fact** is "the quality of being actual, something that has actual existence, an actual occurrence, and a piece of information presented as having objective reality in truth"

A fascinating question is whether there is such a thing as an "objective fact"—whether evolution, or even "more objective" claims, are "true" for all time and for all observers?

What could seem more true than that the shortest distance between two points is a straight line? Yet when it was discovered that space is (barely) curved, this "objective fact" was proven wrong; any "straight" line is infinitesimally longer than one that makes a shortcut through curved space.

So it behooves us to ask ourselves, when considering what we take to be a "fact":

- ◆ Is there such thing as an objective truth?
- ◆ What does it mean that something is a fact?
- ◆ Might the phenomenological school of philosophy be right: that all we observe is only true from our own subjective point of view?

Acid Reign

One of the ways that Daniel Dennett attempts to get around this problem is by calling evolution the "universal acid" that cuts through all competing points of view. The power of this image is that it is just a chemical, and therefore seems inherently objective; acid simply burns all human skin, without regard for human feelings or perceptions. The trouble with this image is that it distorts the nature of human perception—as the product of one's unique mind, where one might be standing, and perhaps, what one just ate for breakfast.

If evolution is indeed a universal acid, this image must be looked at closely. It would seem this would imply that until 1859, and Darwin's book *Origin of Species*, there was no such universal acid that had borne this ultimate, universal truth. Yet what makes this universal? What of other knowledge? What of religion? What of art?

Can every human experience be reduced to the theory of evolution? There are those who would say "yes." There are those who would say "no." There are those who would say "yes" and "no."

The point is it can only be answered by exploring the nature of what we claim we know.

Dennett's mechanistic view of human life may be as much a threat to art as to religion, reducing all "transcendent" human experience to biochemical interactions.

A Brainteaser

Another question to ponder is the following:

If we use our minds to say evolution is true, how can we know that it is true, given that the minds that understand it are a product of evolution? We talked about the problem of a "tautology"—something that only proves itself. But if we are only products of evolution, isn't the most we can know that our brains simply "are," and that to say something is "true" is to introduce a standard that transcends them?

Some Darwinists claim all human thought can be reduced to neurons firing in the brain. Yet how is it the brain can transcend its very process to see that this is what it is doing? Doesn't this perception depend on a perspective that moves beyond the process itself? Doesn't saying something is "true" require a point of view beyond its biochemistry?

Whether or not there is a distinction to be made between the brain and the "mind" is a fascinating question that necessarily blends science and philosophy. The question bears not only on the character of thought, but on the nature of truth itself—giving one to think about the nature of transcendence, and of the possibility of God. If science's greatest strength is its capacity to explain the world by unbroken law, doesn't the act of explaining the world require intelligence that goes beyond it?

Heady questions indeed, but questions that one might believe any man or woman of science would be constantly asking themselves in their quest for knowledge.

Making Predictions

A mainstream critique of Intelligent Design is that it can't make predictions. The same critique is waged against evolution: that it cannot make predictions. Although both ID proponents and neo-Darwinists disagree with this assessment, the truth is in neither case is prediction necessarily a priority.

In the case of ID, inferring intelligence is foremost in its research. In the case of evolution, reconstructing the past to understand how life evolved is the critical program that cannot be tested—insofar as it happened long ago.

Whether or not these theories make predictions, one can imagine ways they would still have value—leaving one to wonder: is predictive power prerequisite to doing science?

Evolutionary biologist Steve Goodwin believes predictive power is overrated.

If the power to predict is not prerequisite to calling something scientific, one must ask: What are the distinctive characteristics that mark scientific inquiry? Moreover, when does "science" become philosophy? Is philosophy ever science? If religion should be barred from the science classroom, what about philosophy?

Nonoverlapping Magisteria?

In the beginning of this book we tried to define the word *science* and what science is. We came to realize that among leading scientists, there is no one accepted definition. We discussed the scientific method, and historic definitions that described the nature of science—specifically, as knowledge that was "verifiable," and later, "falsifiable."

Current understandings argue that science must be limited to material process. We've seen this disputed by proponents of ID, who perceive "transcendent agency" as embedded in the stuff of the material world—of cells, flesh, rocks, and big bang beginnings. Beneath the heated conflicts about Intelligent Design and the process of evolution lie fundamental questions of what science is, and what it is able to tell us.

Stephen Jay Gould attempted to distinguish scientific from religious inquiry. His term *nonoverlapping magisteria*, which is felt by some to dismiss religion, asserts that these disciplines need to be distinct to maintain their own integrities. Yet for others—such as proponents of Intelligent Design—this distinction is not so simple, giving rise to questions about the nature of science, and perhaps of human knowledge itself.

Moving Political Targets

The evolution/ID debate would be well served by open conversations between philosophers and scientists: to the end of exploring the nature of the discipline of science. Until now, most such exchanges appear to be conducted for the purpose of political gain—and lack authentic interest in the other's point of view, in the spirit of enhancing their own. Rather than using strategic argument to maximize political power, academics might remember their commitment to the truth for a larger world.

Proponents of ID need also to respect the mainstream scientific side. There are clearly those who have no underhanded investment in "getting God into the classroom." Yet there are clearly those who do, and who do a grave disservice to this school of thought as well as to the culture. Only mutual trust—in a truth that transcends the factions which comprise the debate—will open up new and creative ways to think about the world, and perhaps God.

Fact or Faith

From departmental gossip, to laboratory funding, to injustices in the tenure process, Intelligent Design has been exploited for the purpose of political power. If ID isn't science, at least it gives faculty the chance to influence the culture in the service of the truth for "the man on the street"—as few issues have in recent years.

God Under a Microscope

One of the most exciting outcomes of the ID debate has to do with its potential to open up new ways to think about God, and reimagine religious faith.

John Polkinghorne, physicist and Anglican priest, eloquently describes ways in which his science has creatively informed his vision of the world, and finally of God. Although most of us will never attain John Polkinghorne's scientific expertise, most people of faith have experienced the presence of God in the natural world.

Yet there is a difference between such inquiry confirming what we already believe and allowing it to lead us to unknown territory where our cherished beliefs might be upended. In Judeo-Christianity, this is the very journey of the wandering Hebrew people; and this is the journey of Jesus, as he wandered the mysterious land of Palestine. If one's faith in God proves to be finally authentic, it will stand the test of the natural world; and if it is not, it will have to be reshaped to be faithful to this world God has created.

As institutionally established as the church has become, Christians are apt to forget that their God was a revolutionary thinker who overturned the religious norms. A radical theologian, Jesus recast that Middle Eastern world's vision of God: defying political, social, cultural, and indeed, theological beliefs. He resisted comfort—in his daily life, and in his relationship with God—in a way that should question the comfortable norms by which we see our own religious faith.

A Unity of Knowledge?

The appearance of design is a favorite theme of Richard Dawkins and evolutionists. They take some pains to grant that design appears to be a feature of the material world; however, for them it is no more than illusion or a product of natural selection. Design may mark a need for order as a way for the brain to organize the world, or perhaps it is a means of giving selective advantage to an organism. For others, such an answer is unsatisfying. Beyond the many questions that remain unanswered by Darwinian evolution, "sensitivity" to a larger, transcendent order seems to persist. Whether seen as religious or artistic or as a simple product of human instinct, the randomness upon which evolution rests seems for them, inadequate.

There may be as many evolutionary explanations as there are explanations of design. But the point may not be to give into survival of the fittest as a moral prescription; the point may be to open up to other points of view to attain the fullest possible truth. Only when we find our human courage behind the cowardice of self-centered pride will we find the power of humility by which great thinking seems to be conducted.

The Least You Need to Know

- In any pursuit of the truth, mutual respect and honest engagement are always more effective than mutual disdain and dishonest charges.

- Intelligent Design is far more than simply "the theory that life is too complicated to be explained by evolution."

- Invoking the word *fact* is often used as a means of silencing counterargument in a debate on a given subject. Yet the nature of a fact must itself be explored.

- The question of whether there is such a thing as objective truth can (and has been) called into question by various philosophical schools.

- Numerous philosophical questions must be considered before coming to a determination one way or the other about Intelligent Design.

Chapter 21

Infinite Regression and Common Ground

In This Chapter

◆ Explaining infinite regression

◆ Is God a certainty?

◆ The supernatural exists

◆ All knowledge is interwoven

In his book *A Brief History of Time*, theoretical physicist Stephen Hawking relates a story that teaches a lesson relevant to Intelligent Design.

"A well known scientist (some say it was Bertrand Russell) once gave a public lecture on astronomy," says Hawking. "He described how the Earth orbits around the sun and how the sun, in turn, orbits around the center of a vast collection of stars called our galaxy.

"At the end of the lecture, a little old lady at the back of the room got up and said: 'What you have told us is rubbish. The world is really a flat plate supported on the back of a giant tortoise.'

"The scientist gave a superior smile before replying, 'What is the tortoise standing on?'

"'You're very clever, young man, very clever,' said the old lady. 'But it's turtles all the way down.'"

Though comical, Hawking's story involves the problem of infinite regression, which stops science in its tracks. If the earth is a flat plate on the back of a tortoise, on what is the tortoise standing? If it is turtles all the way down, what's at the bottom of all the way down?

Infinite Regression's Never-endingness

Scientists try to address infinite regression in numerous ways: often to avoid a beginning, and therefore a Beginner.

> **Evolutionary Revelations**
>
> Celebrated British physicist Stephen William Hawking (born on January 8, 1942, the three hundredth anniversary of the death of Galileo Galilei) is the Lucasian Professor of Mathematics at Cambridge University. It is a post that was once held by Sir Isaac Newton. Though severely physically limited by amyotrophic lateral sclerosis (Lou Gehrig's disease), Hawking is one of the most prolific science writers in the modern world.

As we now know, until the big bang, the "steady-state universe"—in which time and space are a given that has always been and will always be—seemed to make sense as a context in which the rest of life interacted, avoiding the need for any beginning, and therefore any kind of "God."

Hawking's Finite Sphere

Because the big bang newly marked "a beginning," God seemed necessarily implied—unless a way to describe a beginner-less system could be devised.

Thus Hawking envisioned time and space as a finite sphere—with no beginning or end, making it a self-contained, "creator-less" system. Of course the problem remains: if space/time is indeed a boundary-less sphere, aren't we still left with the question: from whence comes this boundary-less sphere? Put in spatial terms, Hawking describes space/time as "finite in extent"—but doesn't calling it finite beg the question of what lies beyond?

Looking Beyond the Beyond

Perhaps answering this question can't be achieved within the bounds of the natural world. However we might try to avoid the metaphysical realm, it would seem that the nature of such speculation requires philosophical answers. Or however we try to reduce the mind to neurons firing in the brain, for many of us such material explanations aren't good enough.

By Einstein's description, science transcends mere chemical processes: "The finest emotion of which we are capable is the mystic emotion. Herein lies the germ of all art and all true science. Anyone to whom this feeling is alien, who is no longer capable of wonderment and lives in a state of fear is a dead man. To know that what is impenetrable for us really exists and manifests itself as the highest wisdom and the most radiant beauty, whose gross forms alone are intelligible to our poor faculties—this knowledge, this feeling … that is the core of the true religious sentiment. In this sense, and in this sense alone, I rank myself among profoundly religious men."

The Certainty of God?

To say we're religious in the way of Einstein is far from defining God. The difficulty that many of us have in attempting to imagine God may be one of the reasons so many have left institutional religion. However one may feel God's presence, to create a *certain* conception of God can leave us doubtful, tongue-tied, confused, even theologically embarrassed.

One of our sources of embarrassment (besides clergy, who imply that it's their job) is most of us assume that certainty is a characteristic of God. It is hard to imagine an uncertain God (for instance, one out of control) because, since the Greeks, God has been generally conceived as unchanging, and certain. But as we shall see, it's not only believers who cling to certainty; many atheists reject such belief because life seems too uncertain.

It is interesting to note that evolutionary theory is also based on certainty. The theory instills adherents to think in mechanistic terms: in which natural selection systematically controls uncontrolled, random mutations. So in both cases—of Intelligent Design and of evolution—certainty seems to be a necessary feature of the theory.

Now we'll look at how theists and atheists are both nourished by certainty. And how and why uncertainty is the enemy of both. Yet it may be that there is another way to understand the nature of God: as utterly open to an uncertain future, by which it bears its fruitfulness.

God Certainly Exists

It would appear that certainty is to God as coldness is to ice. Although many of us doubt God's existence, for those who assent to God, most believe in a God who is certain in an otherwise uncertain world. God is the great "controller in the sky," who knows all that is going to happen; thus God is omniscient through life's accidents, to see through to a determined end.

Fact or Faith _____

Richard Dawkins and Daniel Dennett both presume a certain God in rejecting deluded believers who lack the courage to be atheists. Their conceptions of God appear to be formulated to fail the test of certainty. Unable to allow both certainty and God, they express our own common discomfort.

In the same way, God can answer prayer. In the midst of the uncertainty of illness and random events, God defies life's inherent chance, to be utterly unchanging. Thus, miracles are evidence of God's ultimate certainty—humbling a once-certain natural world by the power of divine majesty.

To say God is subject to uncertainty is to imply that God is weak. Compromised by what he doesn't know, God cannot be ultimate. And if God appears less than the all-seeing deity that we had conceived, it is our lack of vision, not God's, that gives the appearance of uncertainty.

God Certainly Does Not Exist

In the same way, atheism tends to invoke an equally certain God. The difference is that God is rejected for failing this certain promise. For atheists, the unreliability of prayer and of miracles proves, by certainty's litmus test, that God does not exist.

One can especially imagine how frustrating it is to a thinker like Richard Dawkins, who claims his own intellectual fulfillment by the certainty of Darwin. Daniel Dennett's insistence on faith as illusion seems based on a similar assumption that the supernatural must be dethroned by an uncertain natural world. Disappointed by the unfulfilled religious promise of certainty, they may naturally find mechanistic science a more satisfying option.

God Uncertainly Exists

To say that God uncertainly exists seems like an oxymoron. Yet quantum physics disrupted the billiard ball certainty of Newton, asserting the world, at the very bottom, is unpredictable. This is not a matter of inadequate measurement or observation: at the subatomic level, particles simply behave unpredictably.

The location and momentum of electrons cannot be determined at the same time, thereby requiring they be described in a field of probability. And light appears both as a particle and wave, revealing how ambiguity is built into the smallest, most fundamental structures of physical existence. All of which points to an open future which cannot be determined—challenging us to think about God in a dramatically uncertain light.

If the most fundamental building blocks of the universe are uncertain, what does this do to our omniscient God who must see or determine the future? Is God no longer God because he can't predict every billiard ball collision? What of God's relation to the natural world that was supposed to be under his control?

There are no easy answers to these fascinating questions. But as we know, the physical world is critical to Jews and Christians. Rather than seeing science as lying outside of religious faith, descriptions of the physical universe must be a source of religious understanding.

The Incarnation

Unlike some traditions in which the physical world is less significant, Judeo-Christianity takes incarnation seriously. For Jews, the story of God is the story of a human history—of Palestine; of wilderness; of wandering in exile; of Jerusalem; of dust; fatigue; sacrifice; blessings; and joy. For Christians, the story is also embodied in this Hebrew history—but comes to elevate a particular Jew as the incarnation of God.

For Jews and for Christians, the supernatural is alive in the natural world—critically dependent on the physical stuff to bear the spirit of God. Rather than God being abstract, insubstantial, or removed from the muck of life, God is perceived within both traditions as intimately related to the world. In Christianity alone comes the scandalous belief that God became human being, thereby sanctifying the physical world with the possibility of God.

Fact or Faith _____

In the Christian tradition, God indeed becomes man in the human form of his immaculately conceived son, Jesus of Nazareth (also know as Jesus Christ or Jesus the Christ).

Jesus, who was born into a Jewish family, is the savior of the world to Christians, spreading the news of human salvation and then taking on mankind's sins, to die on the cross before rising from death three days later. Jesus is a prophet to Jews and Muslims.

For Christians, the incarnation changes everything. Despite the church's heresy that the physical world is a den of wickedness, God's embodiment in human flesh reminds us of the world's potential. Permeating Jesus, God's spirit is seen as borne by human beings through history—so as part of God's creation, the physical world must dwell in some divine relationship.

A Paradigm Changed

We saw how the ancient philosophers saw design in the universe (despite their insistence on the realm of pure forms as the ultimate domain of truth). And we saw how this tradition was alive in Copernicus, in Galileo, and in Newton. Yet, beginning with Darwin, God was expunged from the scientific enterprise—changing the paradigm, and impeding the relationship between theology and science.

It is this relationship that is pursued and explored by proponents of Intelligent Design. Some may take a deistic point of view, invoking God as a distant presence, whereas others, like William Dembski, assert the world bears a biblical imprint. In either case, the world is explored for the marks of God's creative power, so melding science and theology, and blurring their recent distinction.

Seeking Common Ground

Theistic evolutionists perceive the presence of God—but through autonomous evolution. The difference between theistic evolutionists and most proponents of design is that the latter believe material processes can't explain all complexity. ID looks to mechanisms—such as the bacterial flagellum—and attempts to measure the likelihood they came by divine design.

Most mainstream scientists insist that theology has no place in their work. If God is acknowledged, it is usually a case of nonoverlapping magisteria. Yet Intelligent Design, in noncreationist form (that derives from the ancient Greeks), seeks a unity of knowledge that breaks down the wall dividing theology and science.

Scientific Contributions to Theology

If God is responsible for the universe, it should probably reflect its creator; one should be able to use the physical world as a lens through which to see this God. Although some proponents of Intelligent Design make general claims about God, others are bolder in their assertions—making specific religious claims. We saw how Michael Behe makes minimalist claims that simply deduce intelligence, whereas William Dembski unabashedly invokes a Christian deity.

The big bang is to many theists as Darwin is to many atheists. With the big bang one could be an intellectually fulfilled theist. Until the big bang and its successful predictions, creation made for vague theology; with this sudden revelation, theists could concretely imagine the presence of God.

The biological genetic code, in its immense complexity, has been likened by proponents of Intelligent Design to God's organic language. To them, the cosmological constants that sustain the universe manifest a caring God who enables the miracle of life. So the environment testifies to God's own nurturance—inspiring many of them to perceive these dynamics as divine.

Theological Contributions to Science

Anglican priest Sir John Polkinghorne often receives theologically inspired suggestions for his research. He says he hasn't used a one. Whatever the reason (mainstream scientists might see this as proving their point), it is an important caution to those who would simply meld religion with science.

Polkinghorne is not an ID proponent, but perceives a common motivation. "Theology differs from science in many respects, because of its different subject matter, a personal God who cannot be put to the test in the way that the impersonal physical world can be subjected to experimental enquiry," he says. "Yet science and theology have this in common, that each can be, and should be defended as being investigations of what is …."

Albert Einstein saw science as far more than a mechanistic enterprise—he saw it as a means of human liberation that transcended the material world. Rather than believing science to be distinct from art and religion, he saw them as intertwined in the pursuit of human possibility. "All religions, arts, and sciences are branches of the same tree," Einstein said. "All these aspirations are directed toward ennobling man's life, lifting it from the sphere of mere physical existence and leading the individual toward freedom."

A Unity of Knowledge?

In the face of the knowledge that might be achieved by a multidisciplinary approach—in which science, religion, philosophy, and art move in concert with each other—the academy, and society at large, discourage unities of knowledge, awarding those who carry out narrower kinds of research. Although there are attempts and even examples of fruitful interdisciplinary work, the academic system discourages too much effort in this direction.

The mainstream scientific insistence that Intelligent Design isn't science oddly parallels the ID position that in fact, it is. In both cases, science is treated as the sacred standard for knowledge, and dismisses other disciplines as unworthy of its attention. One might argue that ID shouldn't feel disregarded as "philosophy," just as biology should be explored for its philosophical presuppositions. Indeed, if it's true that there is no crystalline definition of "science," then blurred disciplinary boundaries are in the nature of academic work.

 Blinded By Science

As a university chaplain, I am often surprised at how few faculty know or appreciate the efforts of other faculty: not only in other areas of study, but in their own departments!

A friend of mine in engineering, who heads the driving simulation lab, explained how he would like to write a book for the general public. Yet, however this effort might benefit the American driving public, he explained that such books are "looked down upon" in his discipline as unworthy. Instead, what is valued are articles in the discipline's specialized journals—to be read by a jury of his peers, in the name of tenure and a national reputation.

The trouble with this is the implicit message that knowledge must be specialized—with laserlike focus, and often reduced to indecipherable numbers.

Respectfully Looking Back—and Ahead

We have seen how religion and science were interrelated for thousands of years—in fact, actively nourished each other in their respective points of view—until the appearance of Darwin, and the theory of evolution, when the transcendent was ruled inappropriate to scientific explanations. This wasn't only a consequence of Darwin, but of the Age of Enlightenment, when "reason" was seen to burn through the long shadows cast by the Middle Ages. As has been said, Intelligent Design is therefore not the exception; if anything, non-design science appears to be the anomaly.

And looking ahead, we have also heard the scientific voice of Einstein, whose own deistic point of view permeated his creative physics. However offensive mainstream scientists may find Intelligent Design, they must recognize that it also stands in a distinguished scientific tradition. This is not to say it follows that God must therefore be in the equation; it is simply to say that the inference of God has long been made by scientists.

If the inference of design from the bacterial flagellum seems scientifically "wrong," this should be addressed for its content rather than as mere creationism. Indeed, if any theistic assertion in science appears to be wrong, it should be scientifically disputed instead of dismissed as ignorance. Many fascinating questions go unanswered in the heat of political battle—robbing the culture of opportunities to engage in fruitful learning.

 Blinded By Science

At a recent lecture, I listened to a leading ID critic knowingly lump creationism in with all forms of Intelligent Design. Before several hundred students, this college professor triumphantly ridiculed Intelligent Design scientists. It is such destructive monologue (no questions were allowed after the talk) that does a great disservice to college students, as well as to "students of life."

Mutual Poverty

In the competitive need to publish, get tenure, win grants, and receive awards, the academy may be missing what seems a unique opportunity. Unique not only because such interdisciplinary work is rare—but unique because the eyes of the culture are (for once) focused on its gates. The irresponsible way the debate has been conducted by faculty is leaving behind those who might learn something from them about life.

It would seem that survival of the fittest most applies to the fiercest Darwinians—perhaps proving that if it's the strong who survive, evolution is still "non-progressive." A friend (and adherent of evolution) once said, "Did you ever notice that random mutation seems to apply to everyone but them?" The pride with which many scientists flaunt their expertise defies the Socratic paradox: the wisest know how little they know.

In the same way, ID bulldogs inspire an equally vicious fight: denying scientific evidence, or disregarding it. Their need to score rhetorical points compromises scientific discourse, engaging in debates that reduce participants to two-dimensional combatants. More insidious still is the way this compromises truth by intimidation—whereby those who pursue it retreat to their own subjective and impoverished view.

Toward a Unity of Knowledge

Michael Ruse, an exception to the general practice of irresponsible debate, speaks of evolution as distinct from evolutionism, which he judges to be religion.

> **Evolutionary Revelations**
>
> Why is there a clash? By the beginning of the twentieth century, evolutionism and creationism were competing for space in the hearts and minds of regular folk. It was not a science-versus-religion conflict but a religion-versus-religion—always the bitterest kind.
>
> —Michael Ruse, *The Evolution-Creation Struggle* (Harvard Press, 2005)

The deterioration of evolution and creation into "isms" has resulted in a fight that makes for bad science as well as bad theology. Intelligent Design asks fascinating questions that have been asked for millennia. At its current rate, however, it may never realize fascinating answers, but will continue on as a self-serving conflict, best described by the following remark: "Well, enough about me. Let's talk about you. What do you think about me?"

The trouble is the truth, by its very nature, is never just about oneself. It is always about the "other": other people and a world that exist in relation to the self. People require difference—different people, different points of view, and even different disciplines—to fully realize ourselves, the world we live in, and finally, the greatest truth.

Einstein acknowledged this apparent truth within the rigors of his startling physics. It may also be instructive in the current debate between mainstream science and Intelligent Design. "Science without religion is lame," he said, and "religion without science is blind."

The Least You Need to Know

◆ The problem of "infinite regression" or "turtles all the way down" seems to be unanswerable by the scientific method.

◆ Theists and atheists are similarly wedded to the need for certainty. Uncertainty is the enemy of both.

◆ Certainty is a manifest feature of Intelligent Design and evolution.

◆ Charles Darwin changed the scientific paradigm that had existed for centuries.

◆ A unity of knowledge must be sought in the Intelligent Design debate.

◆ Albert Einstein expressed the importance of the interrelationship between science and religion.

Glossary

abductive inference The making of suppositions from experience by working back from effect to cause, by which ID advocates believe they can show that God is "the best explanation."

abiogenesis The process of going from the primordial chemical soup of the world's beginnings to the complexity of biological life.

allele The heredity "factor" that monk Gregor Mendel revealed as a discrete informational unit rather than a product of infinite blending; opening the way for the "modern synthesis."

amino acids The chemical acids that form functioning proteins in the body when they adopt very specific sequential arrangements encoded by DNA.

anthropic principle Introduced in 1973 by Cambridge astrophysicist Brendon Carter to describe the apparently uncanny way in which the universe seems created for human existence.

argument from "first cause" Thomas Aquinas's causality argument for God, in which he attempts to prove God's existence by asserting that the chain of causes that leads back from the present cannot be infinitely long. Thus there has to be a first cause, which, for Aquinas, is God.

Aristotelian thought Unlike Platonists, who distrusted sensory experience, Aristotelians saw inherent direction in living organisms, concluding that all forms of motion or change have built-in goals or purposes. It was Aristotelian thought that moved cosmologists to mistake the earth as being at the solar system's center.

bacterial flagellum A swimming mechanism that propels bacteria through water, it is imbedded in the membrane of the cell and has more than 40 interactive parts. This is Michael Behe's primary example of "irreducible complexity."

big bang theory The big bang theory asserts that the universe was born some 13 billion years ago in a single fiery explosion. The ongoing expansion of the universe, which the theory predicted, has since been observed by "red shift" light that indicates receding galaxies.

Butler Act The Tennessee law challenged in the Scopes trial, which made it unlawful for public school teachers to teach any theory that denied the story of "the divine creation of man."

Cambrian explosion The sudden appearance of myriad forms of life in the Cambrian period (570 to 500 million years ago), with no apparent ancestors. Evidence for some of "Intelligent Design."

common descent An evolutionary term referring to the idea that all living organisms descend from a common ancestor.

convergent evolution Refers to the emergence of observably similar features in different animals, which, by fossil record evidence, we know evolved separately.

Copernicus An astronomer who, in the sixteenth century, first proposed a sun-centered universe.

cosmology The study of the physical universe, which considers the nature of its creation, as well as its structures and ongoing dynamics, making it a naturally philosophical branch of physics.

creationism The belief that the universe was created by God according to the biblical account. Old earth versus young earth creationism differs in the dating of the creation event: the former, generally holding to prevailing geological estimates; the latter, to between 6 and 10 thousand years ago.

deductive reasoning Reasoning that moves from the general to the particular, and often from the abstract to the concrete.

deism The belief in a God based on human reason—a God that essentially created the world and left it to run on its own without personal, ongoing involvement.

Discovery Institute The Seattle-based think tank for the Intelligent Design movement where research is done, and political strategy developed, for its dissemination in American culture.

DNA Deoxyribonucleic acid contains the genetic instructions that determine biological development. Often referred to as "the blueprint of life," DNA is cellular material bearing the chemical information needed for organisms to grow and live.

The Enlightenment An eighteenth-century European and American movement in which belief in the power of rational thought aptly defined it as, "the Age of Reason." Reacting against the church's longstanding authority, it enabled secularism, and became related to much scientific thought.

entropy The measure of disorder in a given system, the concept of entropy is often used by creationists (and some ID proponents) to question the capacity of strictly material processes to keep the universe from falling apart.

eugenics An outgrowth of Social Darwinism in the early twentieth century, it operated on the assumption that certain racial and ethnic groups were genetically "superior." The eugenics movement gave rise to sterilization laws in 24 states and restricted "unwanted" immigration.

evolution The theory attributed to Darwin that describes "change over time" and strives to explain biological processes as a function of "natural selection."

falsifiability An idea introduced by twentieth-century philosopher of science Karl Popper, to distinguish science from other disciplines of study. Scientific inquiry allows hypotheses to be "falsified"—that is, looks at evidence that could prove or disprove a hypothesis.

fine-tuned universe The term used to describe the seemingly perfectly balanced laws that allow for carbon-based, human life.

First Amendment The provision for "the separation of church and state" guaranteed by the U.S. Constitution, and employed to deny the teaching of Intelligent Design in American public schools.

fossil record The worldwide evidence of fossil impressions that document the history of organisms for hundreds of millions of years. As a concrete account, it is frequently argued about by evolutionists and creationists—though for most ID proponents, these arguments are less important.

Galileo Using mathematics and telescopes, Galileo vindicated the Copernican heliocentric universe—proving that the earth revolved around the sun, and not the other way around.

God of the Gaps A pejorative term referring to the idea that proponents of Intelligent Design and creationists tend to insert God wherever there are gaps in scientific knowledge.

Great Awakening(s) Representing three distinct periods in American history, the Great Awakenings marked fervent religious revival beginning in the eighteenth and extending into the twentieth centuries.

Haldane's dilemma Raised by geneticist J. B. S. Haldane, the dilemma proposes that given the assumptions of evolutionary theory, there wouldn't be enough time for humans to evolve from a so-called common ancestor.

Heisenburg uncertainty principle The founder of quantum physics, Werner Heisenburg found that it was impossible to accurately measure both the location and momentum of individual particles, and that at the particle level, nature can only be probabilistically described, and never defined with absolute certainty.

heliocentric universe A universe, like that defined by our own solar system, in which the sun is at the center.

homology The common genetic relationship between all living organisms, which is generally identified with evolution.

incarnation The belief held by Christians that God became the human form of Jesus Christ, it also refers to any invocation of God as somehow expressed in the material world.

induction Reasoning that moves from the particular to the general, and often from the concrete to the abstract.

Intelligent Design The theory (that proponents contend is scientific) that holds that certain features of the universe are best explained by an intelligent cause rather than by undirected natural processes.

irreducible complexity The idea that there are certain systems that are composed of interrelated parts that contribute to their basic function, but wherein the removal of any of the parts makes such systems break down. An irreducibly complex system cannot be "gradually" evolved because the system's constituent parts only have purpose when they function in concert with the whole.

macroevolution An evolutionary change that moves beyond a given biological species and results in a new species.

memes Richard Dawkins's application of Darwinian biology to social behavior, in which ideas, fashions, tunes, etc. survive and replicate themselves just as genes self-replicate.

microevolution An evolutionary change that happens within a biological species.

Miller-Urey experiment An experiment conducted in the 1950s by graduate student Stanley Miller that attempted to reproduce the process whereby biological life might have evolved from the "chemical soup."

modern synthesis A synthesis that combined Darwin's theory of natural selection with Gregor Mendel's genetic discoveries, making for a mechanism by which natural selection could be scientifically explained.

multiverse A theoretical assertion that our universe is simply one among many. If there are millions of universes, there is no reason to assume that the "success" of ours didn't happen by chance.

mutation A genetic "copying error" by which Darwinists understand the creation of biological diversity. These unpredictable mutations often happen in the presence of radiation, chemicals, and viruses.

natural selection Refers to the biological process, first formulated by Darwin, that results in different traits in organisms that compete to survive and reproduce. When these traits become fixed in organisms, they tend to spread through the population, resulting in novel adaptive traits.

natural theology The branch of theology made famous by William Paley in which God is inferred from perceived design in the universe.

naturalism The philosophical position that all of life can be explained by way of strictly material phenomena.

neo-Darwinism A school of thought that began with the "modern synthesis," combining natural selection with Mendel's genetic discoveries. Evolution could now be understood in terms of "genes," whose competition for survival precipitates evolutionary change.

neo-Platonism A school of thought that held that all physical matter bears an animating life, or "soul." Objects are imbued with active agents that explain motion, growth, and life. Significant to the later Galilean debate about the sun-centered universe, Neo-Platonists were the first to believe the sun was at the cosmic center.

Newtonian physics Newton was known as "the great mechanist," whose physics generally reduced the material world to the interaction of objects. Laws of gravity and motion acted to describe the essential nature of the physical world.

nonoverlapping magisteria Harvard paleontologist Stephen Jay Gould's contention that because of their entirely different natures, science and religion must be kept separate.

nucleotide base The basic constituent of DNA located on the spine of the molecule that serves as the language that informs the building of a protein in a cell.

Origin of Species Charles Darwin's opus work published in 1859, describing organic change through a process he called "natural selection."

paleontology The study of life forms—plants, animals, and other organisms—from prehistoric and historic times by means of the fossil record.

peppered moth An example of microevolution in nineteenth-century "melanic" moths, which purportedly turned dark to adapt to industry-blackened trees in Britain.

phylogenetic tree An evolutionary map tracing organic life from bacteria to human beings.

punctuated equilibrium A theory conceived by Stephen Jay Gould and Niles Eldridge explaining that the absence of transitional forms in the fossil record is due not to incomplete physical evidence, but to real, sudden, punctuated change in the creation of species.

quantum mechanics A theory that deals with physical phenomena at the particle level in more predictive ways than Newtonian mechanics, indicating that there exists inherent uncertainty in the universe.

randomness Often applied to Darwinian evolution, randomness implies that adaptation by Natural Selection works only on errors, and therefore evolution has no overall purpose by which life "progresses."

reverse engineering The process whereby a design theorist attempts to work back from the designed object to speculate about the ways it may have been created.

scientific method The empirical process whereby explanations of natural phenomena are derived from "repeatable," "falsifiable" experimentation.

Scopes trial The 1925 Tennessee court case successfully challenging the Butler Act, which prohibited the teaching of evolution in the public schools.

Second law of thermodynamics Expressed in terms of heat, the second law of thermodynamics says that left alone, a system will always move from relative order to disorder, thereby increasing its "entropy."

Social Darwinism A social theory that uses the laws of Darwinian evolution to explain social behavior. Among such theorists are nineteenth-century philosopher Herbert Spencer, whose concept of "survival of the fittest" became popular with American robber barons, and twentieth-century biologist E. O. Wilson, who effectively reduces social behavior to genetic traits.

speciation The evolutionary change of a population from one species to another, such that biological reproduction between them is no longer possible.

specified complexity Material complexity specifically designed to perform certain physical functions. A junkyard is complex; a Boeing 747 is specifically complex.

survival of the fittest A term coined by Herbert Spencer, following Darwin's introduction of the idea of natural selection, which describes an organism's natural tendency toward survival at all cost.

teleology The notion that there is inherent purpose in the universe, often pointing to a transcendent agent. In philosophy this is called a "teleological" argument. "Teleology" derives from the Greek word *telos*—which literally means "end" or "purpose."

theistic evolutionism The belief that God acts through the biological evolutionary process.

watchmaker The famous "watch analogy" put forth by theologian William Paley describing the way the created world (like a watch) necessarily points to a creator.

weak anthropic principle This asserts that the unique hospitableness of the universe to life means little more than the principle proves itself: the only reason we can be surprised by the universe's hospitality to human life is that we are here to be surprised.

Interview with Michael J. Behe, Ph.D.

Dr. Michael J. Behe, professor of biological sciences at Lehigh University and author of *Darwin's Black Box*, is one of the leading experts on—and proponents of—Intelligent Design.

Here, in an exclusive interview with Christopher Carlisle, author of *The Complete Idiot's Guide to Understanding Intelligent Design*, Dr. Behe discusses the politics of the ongoing debate, why science needs to rethink its approach to ID, and how he came to see design in the universe.

Christopher Carlisle: Describe the relevant personal history that led to your involvement in Intelligent Design.

Dr. Michael Behe: I used to think Darwin's theory was true—pretty conventional scientific beliefs. I'm a Roman Catholic, and we were all taught that God could make life any way he wanted to, and if he wanted to do it by natural laws, and natural selection and so on, then who were we to say otherwise? That always sounded fine to me.

Then in the late 1980s, I read a book called *Evolution, a Theory in Crisis* by Australian geneticist Michael Denton.

In the book, Denton pointed out a number of problems for Darwin's theory that I had never considered, and he presented no alternative explanation: he was just disgusted with Darwinists claiming so much for their theory when he saw a number of problems.

It made me go from simply assuming that Darwin's theory was correct—simply assuming that that's what all intelligent people believed—to asking, "What is the evidence that it claims to explain?"

And when you go from assuming to asking for evidence, it becomes a much less compelling theory in my view. So that's when I started doubting the standard explanation for evolution.

Then in the early 1990s, I kind of serendipitously became acquainted with Philip Johnson. I saw his book, *Darwin on Trial*, when it first came out. I got a copy of it, enjoyed it a lot, and a couple of weeks later I picked up a copy of the *Journal of Science* that had a story on Phil Johnson's book, which I read, because I was wondering what they would think about the arguments that he put forward.

It turned out the story didn't really deal with his arguments: it was just a warning to everybody to keep (their) students away from this book, because it would likely confuse them.

So I wrote a letter to the editor, essentially saying that they should treat his argument seriously, you know, he's an intelligent fellow, and so on. *Science* published the letter, Johnson saw it, and he wrote to me.

From that moment on, I kind of got plugged into the network that he was developing around himself. He's a real schmoozer, and so he was acquainted with a lot of people—whom I didn't know anything about—who also were questioning Darwin's theory. I was then invited to some meetings and events he and others participated in. And from that I discovered I had some ideas that others had not thought of, essentially about biochemical systems and irreducible complexity and Intelligent Design. So by making acquaintances of nonscientists—historians and philosophers—and other people who not only read books, but also wrote them, I got the idea that I could write a book, too. And so I began to think I should write about my ideas regarding Intelligent Design. That's where my book came from.

Carlisle: Because Intelligent Design is perceived in the culture, and particularly in the media, as a kind of cultural brigade or movement, I wonder how engaged you feel in the politics of the movement relative to the science of the movement?

Behe: Well, I'm not a political type of guy. I have my own opinions about things, but I got into science because I wanted to know how the world worked, and so I'm interested in what really explains life, how it came to be. I'm not interested in what some school board does, or other peripheral issues. So I have always seen myself engaged in this topic simply because I want to figure out and essentially let people know what I think is the best explanation for life.

Carlisle: Kenneth Miller [professor of biology, author of several biology textbooks, and a leading critic of Intelligent Design] lumps Intelligent Design in with creationism. In fact, he uses these two terms interchangeably. What do you think about this interchangeable usage?

Behe: I think it's a rhetorical ploy. If you can associate your intellectual opponent with people who are not regarded very well in the intellectual community, then your job is 90 percent done without your ever having to address their argument.

It's kind of like back in the 1950s or so, when a politician accused his opponent of being a Communist if he wanted to increase the minimum wage, or something like that. If he could get the charge to stick, then he wouldn't have to discuss the favorable or unfavorable points of the particular topic. It's the same here. Most people who seriously look at Intelligent Design know right away that it is not creationism, it is not based on a biblical text, it does not require any ex nihilo events. It's very circumspect in what it claims, and so on. Lumping the two terms together is the mark of somebody who isn't intellectually serious about the issue; it's just the mark of somebody who wants to make a debater's point and win in the eyes of the audience.

Carlisle: In your view, is a lot of the resistance to Intelligent Design, within and without the mainstream scientific community, politically motivated?

Behe: Yes. I do think a lot is politically motivated, because some people, when they see Intelligent Design, have visions of Billy Graham entering the White House, or, you know, a flat tax being passed by Congress. They see it as a vanguard of some general conservative political movement, and some people are terrorized of the prospect of that.

And in my view, they really lose touch with reality, and so oppose it, oppose it for those reasons.

I was interviewed on C-Span a couple months ago, and a viewer called in; he was very well spoken and he said, "Well, this is very interesting, but we all know that this is really just a vanguard for a conservative movement that wants to take over, you know, the Congress and the courts."

I couldn't believe my ears. Clearly there are people out there who don't care beans about Darwin's theory or anything else, they just make these free associations, and put ID in the category of—I don't know—a religiously based conservative movement to take over the government and Western civilization as we know it. It's very difficult to speak rationally with such people.

Carlisle: You mentioned your having been raised a Roman Catholic. One of the things I hear a great deal is that a vast majority of proponents of Intelligent Design are Christians, and in particular, fundamentalist, and/or creationist Christians. I was wondering if you would comment on this, as I don't think people appreciate the general Roman Catholic orientation toward science. The implication when you began talking was that the Roman Catholic Church did not in any way influence you to immediately conclude Design.

Behe: That's correct. The Roman Catholic Church's position is very laid back on this issue, compared to other denominations. It does not take Genesis literally. Like I said, it views God as the author of nature, but if God wanted to use natural processes that He made to allow life to occur, well that's perfectly fine. And again, I was taught Darwin's theory in parochial school, in grade school and in high school. So I had no theological reasons to doubt Darwin's theory, or be opposed to it. My own reasons were simply that, as a scientist, I couldn't see how it could account for what it claimed to account for. It simply didn't explain the complexity of life. So a Roman Catholic, in my view, has pretty much unlimited potential to look at the evidence and see where the evidence is leading.

Carlisle: Another conclusion, drawn by your involvement in the Dover, Pennsylvania, (school board) case, is that you want to get God back into the classroom. I was wondering if you could speak to how you perceived your role in that case as an expert witness.

Behe: I was called by the attorneys for the Dover School Board and asked if I would come and explain to the court what is Intelligent Design. So I viewed my role in the trial simply as someone well positioned to explain accurately what Intelligent Design is. I had no part in trying to persuade Dover to do anything one way or another regarding it's biology curriculum. I have never contacted a politician or anybody else, urging them to do anything in particular regarding Intelligent Design. My role, as I see it, is simply trying to explain ideas and trying to push forward this issue intellectually. I am not focused on doing anything in the political sphere, but if anybody comes to me and says, "Well, gee, I would like you to come and explain what is meant by Intelligent Design to this important group, then I see that as part of my role to do so. That's how I view Dover.

Carlisle: Describe your own moment of recognition when mainstream evolution answers appeared inadequate to explain the emergence of biological life. In other words, have you had any experiences in the lab, with the microscope, for example, where you suddenly saw something else than what was explained by Darwinian evolution.

Behe: No, there actually hasn't been any real moment like that.

When you study biochemistry, your basic biochemistry textbook, as a student, you come across all sorts of phenomenally complex biochemical systems. But in my experience, which is typical, your instructor assures you that, yes, we know where they came from: by the process of random mutation and natural selection. And you're a student, you say to yourself, hey, somebody knows all this. I can't figure it out, I don't see all this myself. But somebody has figured out how this came about. So you shrug your shoulders and go on.

I remember studying some systems, like say, the blood clotting cascade, and saying to myself, "Wow. I wonder how this came about by evolution," and then shrugging my shoulders again and saying, "Well, I guess somebody must know," and then going on to the next topic.

Really, the big moment was in fact when I read Denton's book, and I finally realized that nobody knew how these things came about. Everybody was just agreeing and nodding their heads to each other, saying, "Sure we know how this happened," but if push came to shove and you went into the scientific journals and asked what experiment had shown how these could come about by Darwinian processes, it turned out there was nothing there. Everybody was just assuming the answer. And so my own realization was not looking through a microscope, or something I did in my own lab, which is comparatively little relative to what a biochemistry student studies. You study a whole range of things that other people have shown. But rather, it was in realizing that it was not just me that didn't know how these things could arise. Nobody in the scientific community knew.

Carlisle: There have been Intelligent Design books containing point-counterpoint arguments—and the one I'm thinking about may have actually been an article in *Science*, where you put forth the mousetrap metaphor and talked about the bacterial flagellum. And I believe it was Kenneth Miller who responded, and talked about manifest precursors to the flagellum, and so on. So there have been some of these exchanges with mainstream science. Am I right?

Behe: Well, there have been one or two. I think the one you're thinking about was in the magazine of the Natural History Museum in New York—*Natural History*. That wasn't really an exchange. They solicited an article from me, which they gave to Ken Miller and he got a free chance to tear it up and I got no chance to respond. So I wouldn't describe that as point-counterpoint. That's essentially saying, "You stand up here and let somebody take a free shot at your ideas, and you have to be silent after that."

There have been a number of those. Other, better examples are a couple of books in which I contributed a chapter, he contributed a chapter, and a number of other people have contributed chapters on the general topic of evolution and its adequacy, and so on. There was a recent one from Cambridge University Press called *Debating Design* that you may have seen.

Carlisle: Yes, I think Michael Ruse edited that.

Behe: Yes. But there has never been a point-counterpoint in which I got to reply to the points made by the other people.

Carlisle: Why do you think there hasn't been that opportunity?

Behe: I don't know. I suppose it depends on the magazine, but, certainly a magazine like *Natural History* is not interested. They're interested in debunking Intelligent Design. And so they made it clear from the start that there would be no reply. It's either you give an explanation of ID and let the other side swat it, or you just don't participate and we'll swat it anyway. So it's clear that most science journals, maybe even all science journals, don't want a fair exchange—what in any other circles would be regarded as a fair exchange. They are simply interested in, like I say, debunking ID. And as far as other magazines would be concerned, you know, I don't know, maybe they don't consider it of sufficient interest to their readers to set up such an exchange. I'm sure it could go on for quite a number of pages, so I'm not sure.

Carlisle: Would you say that it's been difficult to engage people representing the Darwinian position in a kind of healthy and scientific, as opposed to political, discourse?

Behe: It certainly has been, and it has become increasingly difficult the more publicity that Intelligent Design has received in the past year or so.

As I've said, the problem is that most Darwinists are very antagonistic toward Intelligent Design, so you'd need somebody who wasn't really an evolutionary biologist, or somebody whose science really didn't commit them to Darwinian theory, and those people are less interested in talking about the issue. The ones who are more interested are the ones who see this as a threat, and are interested, not so much in engaging it, as in getting rid of it. Almost from the start that's been the big problem. Darwinists don't want to engage the issues. They simply assume they're right. They assume that questioning of it is illegitimate, and they will not engage in what I consider to be a laid back give and take, you know, just a discussion of the issues. There's always an edge.

Carlisle: A political edge?

Behe: Ah, not just political, but philosophical. Or religious, you might say. A lot of people have definite ideas about the way the world should work, and if you want to call that a theological view, then most people—certainly most Darwinists—have their own views about the way the world should work, and they are loath to put those up to scrutiny.

For example, Ken Miller, in his book *Finding Darwin's God*, makes the case that for there to be freedom in the world, something in Darwinism has to be true. He had the memorable line that God used evolution to set us free. So he has his theological views wrapped up in his understanding of Darwinism, too. So there's that edge.

I think at least as big a factor talking about this with scientists is what I would call political views.

Carlisle: What do you believe Intelligent Design can do that mainstream science can't?

Behe: Well, I think it can evaluate the evidence of biology more fairly, and ask the question, "Can random, undirected processes really account for life?" When Darwin first proposed his view it was seen as radical. It was something nobody had thought of: everybody in the world had thought that life required intelligent direction of some sort to produce it. But now, 150-odd years later, Darwin's theory has not so much been proved, in my view, as simply assumed. And the question of whether or not unintelligent processes can account for life is no longer seen as a legitimate one to ask. Now it's simply assumed that they have to account for life. The question of whether they can has been ruled out—so Intelligent Design can ask that question. If you're a Darwinist you can't ask that question. You already assume its answer. So from my point of view then, an ID proponent has more freedom in that regard than does a Darwinist.

Carlisle: And that would be the primary sort of potential?

Behe: That's right. Does nature explain itself?

Carlisle: Do you believe there are answers that mainstream science can provide that Intelligent Design cannot provide?

Behe: I don't think so. I regard "mainstream" science simply as a subset of what I call free science.

Free science considers whatever answers seem to be indicated by the data, but mainstream science restricts itself to unintelligent answers, or answers that do not point beyond nature. So there is nothing that mainstream science could do that a freer science—which I consider Intelligent Design to be—cannot do. I mean, people who are working on an Intelligent Design view can say to themselves, well maybe nature doesn't show any signs of having been affected by an intelligent agent, or having been set up by an intelligent agent, but maybe it does. Let's look and see. And we can come to the conclusion that no it doesn't. But mainstream science starts by begging the question, by assuming that, no, nature does not point beyond itself, so under that assumption, let's come up with the best answers we can. So I regard mainstream science as restricting itself, where Intelligent Design is not restricting itself, so there's nothing that mainstream science can do that ID could not.

Carlisle: You've often said that Intelligent Design makes minimalist claims: particularly with respect to theological inference. I'm just wondering about that. When you say—if I'm getting it right, and if not, please correct me—that Intelligent Design will always be unable to make those theological inferences, or identify the agent, do you believe it's a matter of time?

Behe: Well, you know I don't say that Intelligent Design would never be able to identify a designer or some such thing. You know, suppose that Francis Crick was right, and that aliens made life on Earth. Who knows, maybe if they came back for a visit we would be able to meet those aliens. I regard that as unlikely, but the minimalist Intelligent Design claim is that some intelligent agent designed life, and it could be a supernatural being, it could be some smart space alien, or something else.

If, in fact, the intelligent agent is a supernatural being, as I suppose, and as probably most people do, then it is entirely possible that science would not be able to identify the designer. But science is not the only intellectual discipline we've got. There's also philosophy and theology, and just because science can't identify something doesn't mean that we are left with no resources to try and answer the question. And I would probably just say that if that is the case, if the designer is some being beyond nature, then science won't be able to answer the question. But perhaps philosophy or theology can.

Carlisle: Often in this debate you hear critics, largely from mainstream science, talking about the incredible complexity of science, and the fact that laypeople can't understand the science and can't understand the issues. What I'm wondering is: a) how do you feel about this attitude; and b) if you think that's not entirely correct, what is the role of nonscientists in the debate?

Behe: I think people can understand the issues if they're presented clearly and so on, just as laypeople can understand any other question in science. Unfortunately, in my point of view you run into a problem because some people really want to purposely muddle or obscure the issues.

For example, when somebody like Ken Miller comes out and talks about these guys as being creationists, and they're just trying to, you know, get the Bible into public schools, etc., he [Miller] can do it easily by using a lot of examples, or by pointing to a lot of things that I think mislead or obscure the points. It's easy for people, for scientists, to essentially blow smoke around the issue where a nonscientist can then become confused, even though the issues themselves are pretty clear and can be presented clearly to a lay audience.

The problem is when you have a technical person coming in who is determined to obscure those issues. So what is the role the lay public plays in the debate? I don't view the role of the public as participating in a debate. I just view the role of science as trying to tell the public, in its best estimation, how the world works. So the public is interested in knowing what the world, the universe, is like around them, and science's job is to honestly portray to the public its best estimation of the answer of a given question.

Unfortunately, it gets pretty convoluted, because a lot of scientists have definite prejudices about the way they want the world to be, and they will sometimes filter the results of science through the lens of their own philosophy or preferences in describing the universe to others.

In discussing these issues, a number of scientists have quite forthrightly said that science has to restrict itself to natural explanations. Unfortunately, they seem to confuse that restriction with a restriction on reality. And they will tell the public that science knows that natural explanations can account for life and all of nature. So they take a restriction on science and impose that restriction on the general public as well, and they take—what some call—a methodological restriction on science, and they mistake it for the way that the universe really has to work. That's a real problem, and the public then gets informed of a view of the way the world works and they're not told that it's being filtered through some sort of philosophical lens before it gets to them. Consequently, the public gets a skewed view of what the evidence really shows.

Carlisle: So how do you feel about that methodological restriction within the context of science itself?

Behe: I think it's stupid, quite frankly. I think science should be a pursuit of the best explanation for the data, using physical data that's available to everybody. So you're not relying on secret revelations or something like that, or you're not relying on sacred texts or anything. If you have a laboratory result, and you use standard methods or standard logic, standard methods of reasoning, and if it seems to point beyond nature itself, if the results of nature seem to point beyond nature, well then, that's the way they seem to point.

I see no intellectual virtue in pretending that the data says something other than what it seems to be saying. And the only time when the idea of methodological naturalism is evoked is when the data seems to point beyond itself in debates like over Intelligent Design. It seems to be an intellectual gerrymander simply to keep explanations that seem to have religious overtones out of science, but it's not aimed at getting the best explanations for the data, and I think it's an artificial and very unhelpful rule.

Carlisle: Would the bacterial flagellum discussion be an example of this?

Behe: Yes, I think so. When I make the case that the bacterial flagellum strongly points to design, a lot of people will say well, we don't know how the flagellum could have arisen by random mutation and natural selection, but science simply has to think that there is some natural explanation out there, so we'll keep looking.

To my mind, it is a denial of a fairly obvious explanation simply because the data is pointing in a direction some don't like. They don't want to go there. So they're ignoring reality and essentially closing their eyes and hoping this will go away, and in the future maybe they'll come up with an answer that is more to their liking. But if science ignores the data in the hopes of something more palatable coming along later, then science will run into trouble pretty quickly.

Carlisle: Am I right that Kenneth Miller's counterargument is there are precursors leading to the creation of the flagellum, and that essentially there were parts that were used for other things that eventually got used for the flagellum?

Behe: Well, it's hard to figure out exactly what his argument is. I think his pointing to the type-3 secretory system, which shares components, or shares things that look like components, with the flagellum, is simply to say that the definition of irreducible complexity doesn't work, that you can use a subset of the flagellum for some other task. And he claims, misleadingly I say, that the definition of irreducible complexity means that you can't use any part of an irreducibly complex system for another task. That's incorrect. That's not what irreducible complexity means.

Leaving that aside, he is not even trying to explain where this type-3 secretory system came from, how it changed into the flagellum, or any such thing. It's simply an attack on the concept of irreducible complexity. So at that point he stops and says, well, irreducible complexity is wrong, so ha ha, and the assumption is, well, I guess Darwinian evolution, or something like that, must work. But he never even tries—certainly not seriously—to describe how undirected processes could account for the flagellum: it's simply a negative argument against irreducible complexity. So no: he doesn't try to explain how this could be a Darwinian intermediate, it's essentially just an argument against my argument.

Carlisle: You alluded to Miller's misunderstanding of irreducible complexity. Could you describe what you think his problem is with irreducible complexity, especially with respect to his use of counter—what he considers to be counter—examples of precursors?

Behe: Irreducible complexity says that if you take a piece away, then the function of the system that has all of its parts, is lost. It does not say that some pieces of the now broken system can't be used for other things or have no function: it's that the function of the entire assembly is gone.

For example, if you take away critical pieces of the flagellum, it might now work. It might still be able to work as a pump, but it sure can't work as a rotary motor, it sure can't act as a propulsion device for the bacterium.

With the mousetrap, he takes out a piece and says, well, I can use a part of this as a toothpick.

Yes, well, he can take some parts away and he can use them to pick his teeth, but he can't use them as a mousetrap. If you take away the spring, or any other part, the function of the system is broken. So essentially he's just trying to again attack the concept of irreducible complexity. But if you took a real toothpick, you know—he also uses things like a paperweight or a key chain as other functions, other purposes one could use parts of the mousetrap for—well, you could use it as anything. You could use it as a doorstop, you could use it as a hat, if you wanted to, but if you took a real hat, or a doorstop, or a key chain, none of those things are on their way to becoming mousetraps, so it is not an intellectually serious way to really show how an unintelligent process could produce a mousetrap. It is simply, in my view, a rhetorical trick to try to get audiences to think there's something wrong with this idea of irreducible complexity. It's not the mark of somebody who really wants to understand these problems. It's the mark of somebody who just wants to be rid of the idea of Intelligent Design.

Carlisle: What do you feel are the most destructive misconceptions in the culture at large about Intelligent Design? Given your opportunity to read the culture, and read the responses of the culture to both sides of the debate, are there certain misconceptions about Intelligent Design that are particularly troubling to you?

Behe: Well, the biggest misconception, in my view, is that this is essentially wishful religion thinking. That it is motivated by a desire to read God into nature, and in my view, that's completely opposite to what it is.

It is that—unexpectedly, in different branches of science over the past hundred years or more—startling and unexpected evidence has been found that points strongly beyond nature: that nature, what we see in our universe, does not explain itself. And the start of this was not in biology. It was in physics where scientists, much to their surprise and chagrin, found that many of the features of the universe seemed to be finely tuned, just so, to permit life to exist in the universe, and that the more and more we know about the universe, the more and more we see that this fine-tuning is everywhere you look.

In astronomy you discover that a planet can't exist just anywhere, it's got to be in the right region of the solar system, and the star has to be in the right region of the galaxy, and the planet on which life is to occur has to be the right size and have the right mix of elements and all sorts of things. In biology, a hundred years ago, people thought the cell would turn out to be a little blob of jelly, pretty simple, and much to the surprise and chagrin of scientists, a cell has turned out to be stunningly complex. Again, exactly what you'd expect, pointing to unfathomable technology, really, really smart design.

So I think the really pernicious idea is that Intelligent Design is reading into nature what one wishes, where I view it as a surprising, unexpected reading of what nature is actually trying to tell us.

We are getting results that nobody foretold, and mainstream science is trying to shoehorn them into an old view of a self-explaining universe when I think nature is trying to tell us something different.

Carlisle: What would you most like to say to your mainstream critics to effectively maximize the communication of your position? Is there something distilled that comes to mind?

Behe: Well, it's hard, because many of my critics have become, in my mind, pretty hardened to all this. But I would say that—believe it or not, critics—I am just honestly trying to account for this physical data. I have no particular agenda that I am trying to push forward. I think science should be a no-holds-barred search for the truth. It shouldn't restrict itself to one set of answers, even the answers. It shouldn't restrict itself to saying nature has to explain itself if in fact the data of nature pointed in a different direction.

Interview with Michael Ruse, Ph.D.

Dr. Michael Ruse, professor of philosophy at Florida State University and author of *The Evolution-Creation Struggle*, is considered an expert in the philosophy of science (specifically the social impact of Darwinian theory) and one of the leading opponents of Intelligent Design.

Here, in two exclusive interviews with Christopher Carlisle, author of *The Complete Idiot's Guide to Understanding Intelligent Design*, Dr. Ruse discusses his concerns about the ongoing debate, what he believes are the underlying motives behind ID, and why he believes ID is not—and should never be considered—science.

Interview 1

This is based on e-mail correspondence between author Christopher Carlisle and Dr. Michael Ruse.

Christopher Carlisle: You say, "Nothing makes sense except in the light of evolution." What do you mean by this?

Dr. Michael Ruse: I don't mean this in a particularly profound or mystical way—simply that in order to understand the present, usually if not always, the secret is to find out the past. For instance, if I wanted to discover what sort of person you are, my first set of questions would be about your past, your parents, your training, and so forth. Then I would know things like whether you are a priest because of family influences or if because of a Saul on the road to Damascus experience, and already I would be on the way to understanding you. If you come from a long generational line of rich Episcopalians that is one thing; if you come from poor Catholics, that also starts to tell me something about you.

Carlisle: Describe your principal concerns about Intelligent Design as a mode of inquiry into the nature of the world and biological life.

Ruse: I just don't think that ID is science—it is natural theology dressed up in the guise of modern science. In one way, I have nothing against someone being keen on natural theology—although if I were a believer I would be more drawn to the thinking of someone like [Karl] Barth who did not care for natural theology—but I don't think it should pretend to be science. I think ID has an agenda, and that it is profoundly anti-Enlightenment and that does worry me for moral as well as epistemological reasons.

Carlisle: Do you believe there is a clear distinction between science and other modes of inquiry (like philosophy and theology)? If so, what is it?

Ruse: Yes I do and I have been saying this for years, notably in my testimony in the Arkansas trial in 1981. I think science works by understanding through unbroken law, and that religion does not. Science is therefore an empirical inquiry right down the line. I am not saying that religion is false, but it is not science. If Jesus rose from the dead that may be true, but it is not a scientific claim—even if you think that no laws were broken and the rising was a spiritual matter, inasmuch as you are claiming that it is true, this is not a claim made in terms of laws. Philosophy is more difficult. Many would say that philosophy is not science. I am more of a naturalist on these matters and would say that one should aim to make philosophy into a science. So when I try to (for instance) look at the natural of scientific revolutions, I look at the science and its history and try to draw generalizations from that and not to proceed a priori.

Carlisle: Are there any ways that you believe Intelligent Design has made a positive contribution to current knowledge?

Ruse: No—I suppose it has had the virtue of making people like me explain more clearly why we are Darwinians. That is the purpose of my new book (July) *Darwinism and Its Discontents*. But this is all in a negative sense.

Carlisle: The Intelligent Design debate is steeped in politics most often depicted (by the media and the academy, in particular) to be right wing, religiously motivated partisanship. Do you believe this is an accurate depiction? Can you imagine Intelligent Design proponents who are not so motivated? Do you know any?

Ruse: Well I do agree that politics are involved and this is the theme of my last book, *The Evolution-Creation Struggle*. I think the ID people have a conservative, anti-Enlightenment (anti Post Millennial) program and that it comes out quickly in the writings of people like Phillip Johnson. I do not know, and frankly would be surprised to find, any ID enthusiast who did not think this way.

Carlisle: In your book *The Evolution-Creation Struggle*, you speak of "evolutionism" as a kind of religion. Would you explain this contention?

Ruse: I think that many evolutionists make more of the theory than just science—they want to surround it with moral prescriptions and look upon it as something telling of humans' special place in the creation and argue that if you are an evolutionist—a Darwinian particularly—then you must be at least an agnostic and preferably an atheist. ("Must be" in the sense of "should be"). Thomas Henry Huxley was a prime example in the past. Edward O Wilson is a prime example in the present.

Carlisle: You have been critical of some evolution proponents as participants in the Intelligent Design debate. What has bothered you most?

Ruse: I worry that people like Dawkins and Dennett are doing more harm than good. They hate all Christians and other believers and so alienate many people who are very keen to put down ID. More than this, their strident atheism and contempt for believers—like Dennett's term "bright" for those who agree with him—puts off many people in the middle, who are sincere Christians but do not necessarily have a strong axe to grind to support ID. Not all Middle Americans are right-wing Americans.

Carlisle: What concerns you most about the debate from the side of Intelligent Design?

Ruse: I don't really have an answer to this one, because I am not from the side of ID. I think that folks like Dennett and Dawkins are insensitive to the feelings and thinking of many decent people, but this does not excuse ID.

Carlisle: Do you believe the Intelligent Design debate is worth having? If so, how does it need to be conducted?

Ruse: No. Of course we who think that crude evangelical religion is a bad thing and ought not to influence the public debate have to have it, given the existence of ID, but this is the only reason. There is no intellectual merit to the debate. Having said this, my feeling is that those of us who are anti-ID should be formulating our own moral positions, we should be working together and not quarreling among ourselves and above all we should be trying to formulate a position on the science-religion relationship that shows how science and religion can coexist. (This last task is just what I am setting about myself in my next book.)

Interview 2

This is based on a telephone conversation between author Christopher Carlisle and Dr. Michael Ruse.

Christopher Carlisle: In response to your written comments, my first question is, what do you mean by the Intelligent Design movement being "anti-Enlightenment"?

Dr. Michael Ruse: Well, I mean that I look at anti-Enlightenment as implying literalism, moralism with respect to reading the Bible. I don't think you necessarily have to be an atheist, but I do think you have to be prepared to move away from the literal readings of the Bible and such things.

I take it that Enlightenment means thinking that science is extremely important. Again, I don't think you have to believe that science can necessarily solve every problem, or that there are no problems that are not scientific, but I do think you have to give science a very important and rather dominant role in your world picture.

I take it that anti- or that Enlightenment (thinking) means that you're going to be more tolerant of people's foibles. And I would add that, obviously, I think that Creationism, as such, is clearly anti-Enlightenment, with biblical literalism and the emphasis on "an eye for an eye," and "a tooth for a tooth," and all those sorts of things. And as you know, I want to say that although obviously Intelligent Design theory is not necessarily crude biblical literalism, I think it's very much a case of being "Creationism Light" in the sense of being part and parcel of the same overall thrust of wanting to circumscribe the power of science. And as I said, I think that the ID people have a political and moral agenda, and that you see this particularly in the writings of people like Philip Johnson, where he's very much against things like gay marriage, or anti-capital punishment, or things of that nature.

Carlisle: One of the things that I find interesting is that Michael Behe contends that, first, he is not even politically interested, and second, he is not really even interested in the kinds of large animal evolutionary issues that someone like Richard Dawkins is. Behe is really just interested in biochemical systems, which are far smaller, and he would argue more "nonpolitical" in nature. What do you think of that?

Ruse: Well, obviously there's a certain truth to these sorts of things. I mean, when Michael Behe's writing *Darwin's Black Box*, he's basically talking about things like irreducible complexity and stuff like that, and he's not spending a lot of time on the moral issues.

On the other hand, I think that this is all part and parcel of a kind of package deal. I'm not saying that I think all ID people agree on the same thing. Clearly they don't. Paul Nelson, for instance, is a "young-Earth" creationist; Michael Behe obviously accepts a great deal of evolution. So as I say, I don't think you can say that everybody believes exactly the same thing, but I do think that they are united by a similar overall set of concerns. I think it's really significant, for instance, that Michael Behe and his wife home school. I mean, you know and I know that, in this day and age, that home school means either you're way to the political right or you're way to the political left. You're either, you know, anti-abortion, pro-capital punishment, and all of these things, or you're into sort of, horrible sandals and tying ribbons on trees.

By and large, Episcopalians don't home school.

Carlisle: No, I would say that's generally ….

Ruse: I mean, they might send their kids away to boarding school, but they sure as hell … they're not big on home schooling.

Carlisle: Okay, you say that ID has an agenda and that it is profoundly anti-Enlightenment, and you say, "That does worry me for moral as well as epistemological reasons." What are these moral and epistemological reasons?

Ruse: I mean, look, I know Phillip Johnson gives the game away.

Johnson says again and again and again, the big issue here is naturalism. He says—I think you'll find the others echoing him in this—he says what we've really got here, it's not so much about gaps in the fossil record. I mean, you know, Johnson, for instance, says quite openly, he says, "archiopterips," score one for the evolutionists. I mean, he's quite, you know, as I say, Behe is prepared to allow huge parts of evolution. And I think probably Bill Dembski is, too.

So as I say, I don't think that's what's keeping them awake at night, but Johnson is quite open about this. He says, the issue really is naturalism. He says, of course, he knows the game that evolutionists and others play, he says, "Of course, you folks say,

'Well, we're into methodological naturalism, but not metaphysical naturalism.'" In other words, we're, as scientists, happy to play the let's pretend we're all atheists trick on Monday through Friday, but it doesn't imply what we're going to say on Sunday, you know, whether or not there really is a God. But Johnson said the case is that the trouble is methodological naturalism and metaphysical naturalism in actual fact are usually pretty close together and so consequently, people who are pushing evolution are in fact part and parcel of an overall, what do you call it … philosophy, metaphysic, or something like that. And that would apply both to liberal Christians, as well as to agnostics, as well as to atheists. So he'd put us all in the same camp. And what he wants to say is that he's not a naturalist, he's a theist. And of course, to him theism means this rather, you know, as I say, this right-wing reading of the Bible, you know, "an eye for an eye, a tooth for a tooth," capital punishment. I don't know where they get abortion out of it, but you know what I mean.

This whole, sort of … and the anti-homosexuality, because Leviticus says it and St. Paul says it. And as I say, for Phillip Johnson, his theism means those things, whereas he feels that naturalism is something that says homosexuality is not a, you know, a choice, it's not a matter of original sin. It's just a matter of the genes or the upbringing, and basically somebody's either gay or straight and there's not a hell of a lot you can do about it, but certainly one thing you should not do about it is make moral judgments on this issue.

If somebody is gay, that person can be just as good and worthwhile a person in the eyes of the Lord—even if they're not only gay, but practicing gay, as somebody who's straight. I really do think that these are the sorts of issues that we've got at stake here.

Carlisle: Do you believe naturalism and theism need to be at odds?

Ruse: It depends on how you define it. I mean, the way that Johnson defines it, they're almost by definition at odds. There's no way that you can follow a Johnsonian program and be a naturalist of any kind.

Having said that, I've long argued, of course, that you can actually be a Christian. Take naturalism, certainly take evolution, certainly take science, an awful long way, and never the less be a very, you know … a good, "small c" conservative Christian. By "small c" conservative, I mean one who takes seriously the notion that Jesus died on the cross for our sins, and rose on the third day and things of this nature, which I take to be, "small c" conservative; what I mean is, I don't think you have to dilute it to such an extent that a Unitarian can say, "Well I can accept all of that."

You know how it is: people can be very different. I mean, we see this in people like Jefferson, people can mean very different things by what they mean by—when they say that they're "Christian." I certainly meet people who say well, "I'm a Christian, but I deny the divinity of Christ." Now, personally, I like to say that I'm a conservative nonbeliever. I don't think that you should call yourself a Christian if you deny the divinity of Christ, I don't care what a good bloke you think he is.

Carlisle: Yes, I think it's important to be clear in defining.

Ruse: I mean, personally, I think—when I say "small c" conservative Christian, I think that a "small c" conservative Christian has to think that God is creator of heaven and Earth out of nothing, has to believe that humans have a special place on this Earth. We might not be the most important, but we have a special place and we are made—in some sense—in the image of God, that we are sinful, and that Jesus came to save us. I mean, I see these as bottom-line claims for a Christian. And I think that you can accept all of those things and still be a Darwinian.

Carlisle: Right.

Ruse: I'm with John Henry Newman [nineteenth-century Anglican priest turned Roman Catholic cardinal] on this. You know, I mean, I am not a believer, but my God is the God of love and mercy, the God who died on the cross for my sins and made possible eternal salvation. My God, you know … how my God produces individual humans, or individual species has nothing to do with that.

Carlisle: Okay, I think you make the case that creationists, Intelligent Design proponents, and evolutionists, all inappropriately make moral prescriptions. I'm wondering, first of all, if that is the case, and if it is the case, in your view are there any appropriate moral prescriptions to be made?

Ruse: I mean, I don't think that the moral prescriptions in themselves are necessarily inappropriate. I'm writing a blurb for Ed [E. O.] Wilson's new forthcoming book, that once again wants to push biodiversity and things of this nature, and I for one am all in favor of pushing biodiversity. What I think is wrong is to try to do this in the name of evolution. I think what is right is to use evolution to understand how one might achieve one's ends. It seems to me that anybody who is concerned, for instance, in the modern crisis of global warming and all of these things is bound to take seriously what we call modern science and ecology, for instance. But my point is I don't think that the moral prescriptions come from hydrology, or from ecology or anything like that. And where I would fault that is, I think he wants actually to get his moral prescriptions out of his science, because he thinks that evolution is, in some sense, inherently progressive and would therefore, this being a value-laden notion, mean that it's laid upon us morally to promote the evolutionary process, particularly inasmuch as it's linked to humankind.

Carlisle: But on the other hand, you have someone like Richard Dawkins who desperately wants to keep teleology out of the program.

Ruse: I have written a little piece for [a Festschrift for Dawkins just published in England] what Dawkins first wrote, but basically I was being nice—basically I was trying to suggest that Dawkins is pretty incoherent on what he wants out of all of this. On the other hand, Dawkins does know the moves that you can't get out from this, but on the other hand, he does want progress of some kind and he certainly thinks that evolution is something that denies Christianity and so on and so forth. So to a certain extent I think that Dawkins runs with the hare and runs with the hound.

Carlisle: But I had the impression that Dawkins did not want to have any kind of scientific progress.

Ruse: Dawkins is really keen on progress. There's no question that Dawkins believes not only in scientific progress, but Dawkins believes in biological progress. I mean, he's said that again and again and again and again. The question is to what extent does Dawkins believe that this is a value-laden notion, and he wants to deny that it's that. He wants to say that it's a purely objective measure. Rather like, you might want to say that there are more books in the Library of Congress than there are in the Yale University library, let's say, and therefore, on a scale of, let's say a large number of books, that the Library of Congress beats Yale, beats FSU. On some level that's what Dawkins wants to get, whereas Ed Wilson wants to go one step further, and say, therefore, the Library of Congress is better than Yale, is better than FSU. And Ed, I think, thinks he can get that straight out of the science. I want to be fair to Dawkins. I think Dawkins recognizes that one can't do that. But then, somehow it seems to me that Dawkins wants to—even though he's not going to make those moves—to a certain extent he wants to bask in the glory of having made them.

Carlisle: Right. Okay—shifting gears a little bit—how would you define "natural theology" over and against "modern science"?

Ruse: Natural theology, I take it, is the attempt to prove the existence of God through reason and evidence, so natural theology would encompass the ontological argument, the cosmological argument, the philological argument, the argument for miracles, those sorts of things. I mean, this has a long and variable history and I think it was pretty successful up to the eighteenth and nineteenth century when, of course, people like David Hume started to hammer on it pretty heavily, philosophically. Then of course, along comes Darwin.

Carlisle: Who upends natural theology?

Ruse: My personal feeling is that today I would want to say I really don't think natural theology works. I don't think it's a stupid thing. I don't think it's bad, but I just think it's something that we now see just doesn't work. Now, there are theologians that don't agree with this.

It seems to me that somebody like John Polkinghorne, with his enthusiasm for the anthropic principle, add on Paul Davis and these other people who are so keen on the anthropic principle, who always say, "No, we're not in favor of nature theology." Well, it seems to me that they're getting pretty damn close to it. What they're saying is: the evidence of science is that if the world's continents weren't exactly as they are, nothing would work, so that can't be chance, da-dum, da-dum. There must be some being over and above.

Now I'm much more inclined to go—and it's a theological position, as you well know—I'm much more inclined to go with the sort of tradition which you find in [Søren] Kierkegaard, and certainly find in Barth, and all that group known today as the neo-orthodox group who want to say natural theology just doesn't work, and even if it did, there's something wrong with it.

Again, I come back to Newman. The whole point, Newman says, "I believe in Design because I believe in God, I don't believe in God because I believe in Design."

I think that there's something right in Kierkegaard's insight that if you can prove God through natural theology, at some level it devalues faith, because faith must be some kind of existential leap into the absurd—not stupid, but into the absurd.

In other words, faith requires that kind of—I mean, an almost Billy Graham kind of commitment. Graham doesn't get up and say, "I'm going to give you proof of the existence of God." Graham says, "Believe." And although I'm not prepared to fight with Billy on this, I don't think he's being theologically wrong about it. It requires some kind of commitment. [Wolhart] Pannenberg uses the phrase "theology of nature" rather than "natural theology."

It seems to me that if you're a Christian, say, you're surely going to see the world as filled with the glory of God, but you're not going to say, "This proves God." You're going to say, "As a Christian, I see the world through—if you like—my Christian filter. And therefore, to me, the world is that much richer and more worthwhile and more significant."

Carlisle: Would you say, therefore, that the move to "prove God" is in itself problematic?

Ruse: Yes, I would, because that's natural theology to me. I think that this is part of what, you know, part of what is motivating Intelligent Design, of course.

Carlisle: The need to prove ….

Ruse: If you said to me, Intelligent Design is just refurbished natural theology, I don't think that would be a mistake, because you have to put it in a cultural context. I don't think that Archdeacon Paley was, in his day, pushing a conservative moral position. He was utilitarian, for goodness' sake. Whereas I see what these people are doing, what's motivating them is not trying to show that Karl Barth is wrong, but rather to promote these social values.

Carlisle: So your position would be different from Dawkins on this point, because Dawkins says, "Paley was wrong, gloriously wrong."

Ruse: Yes. Now, I'm inclined to agree with Dawkins on that. I mean, I'm a Darwinian, so the point that Dawkins is making, and I think all Darwinians agree on this, as opposed to, say, Stephen Jay Gould, is to say the key interest, the important issue in the organic world is the designlike nature. And natural selection explains that. And that is, of course, what Dawkins means by Paley being gloriously wrong.

Don't forget that the person that Dawkins has in mind there is not your Christian believer, really. He's got in mind someone like Stephen Jay Gould, because the point he's making there is not that God doesn't exist, I mean he believes that, but the point he's making is that Design is something that needs explaining. It's just that you can't do it through God, you do it through natural selection. Whereas somebody like Gould might say, "Well I personally," speaking as Gould, "I personally think the whole Design issue has been overdone."

Carlisle: Would you elaborate on your contention that science works through unbroken law, and theology does not?

Ruse: It's not a question of theology does not. Let me just simply say that science since the Enlightenment has moved increasingly rapidly to the point that when you get to the nineteenth century, pretty much exclusively science is something that works by explaining through unbroken law. I mean, its been successful, so we got to the point today where science is simply going to say, "And that's the way it is," that there's good evidence in the past, where even if you don't have solutions, it's better not to give up, it's better to keep going and that's what we're going to do. That's the scientific method, the scientific attitude. So I would want to say that's what science is all about. Now, I don't think religion is even in this business. Religion, I would want to say, is trying to talk about issues to do with meaning and things like that.

Now it may be that the religious person, because of his faith, wants to impose what is known traditionally as the order of grace on the order of creature, and this religious person might say, yes, inasmuch as science tells you anything, water did not turn into wine; however, as part of my faith commitment, I think it did. I'm not making a scientific claim now, I'm making a religious claim.

This is why I am personally always very uncomfortable with people like Pannenberg who are trying to prove the stone really did move. It seems to me that's completely misdirected. I mean, maybe the stone did, maybe it didn't.

Frankly, I don't think it matters a f---. I mean ... the point is, I don't think you're going to prove it. I just don't think you're going to prove it in the face of science. So if you want to believe that it really truly did, all you can do in the face of this is to say, I have faith that Jesus is my Lord, that he rose on the third day, he really did, and therefore the stone must have given way. But this is an entirely faith-based statement, and I'm not even going to get into science on this. But these people who spend all of their time trying to discuss whether or not women at that time and age would've been reliable witnesses, and if they weren't then why were they taken seriously, I mean, come on, give me a break. These things aren't important.

Carlisle: So do you think there's an absolute disjunction between the dynamics of the material world and the dynamics of the theological, you know, the faith?

Ruse: There can be a disjunction in the sense that we are physical beings in the world. But if you are theological, then you believe that we are not just beings, but beings, at some level, made in the image of God.

So obviously, it's rather like the mind-brain thing. I think at some level the two come together. But I just don't think that the way to do religion ... I mean if you try to get into competing with science you're going to lose all the way.

I think that where religion has to work is in those very areas where science cannot speak, you know? Like ultimate origins. It seems to me that science has adopted a kind of machine metaphor, and that is, basically, you start with the machine. I don't think that religion can speak to that. And I don't think science can speak to ultimate moral standards. I don't think science can speak to, is there a, you know, a world beyond ours? I think there's all these sorts of places where religion can enter very meaningfully—I'm not even sure that science can speak to consciousness. It seems to me that if you adopt the machine metaphor, machines just aren't conscious. End of argument. I would be inclined to say that not all dimensions of experience, if you like, which religion can speak, but these are not scientific understanding. It doesn't go to say that you can't get scientific understanding, it's like the meaning of, let's say, the Saint Matthew's Passion, something to which I am absolutely devoted. I mean you can give all sorts of scientific arguments, musicological arguments about these things, but ultimately I think I could just say that Bach speaks to me in some spiritual way, or something like that. What I am trying to say here is that I get a kind of understanding of human nature by listening to the Saint Matthew's Passion that I don't think I could ever get from science.

Carlisle: Now, I believe you said, there is no intellectual merit to the Intelligent Design debate?

Ruse: There is. Intelligent Design is old-fashioned natural theology, but it's being promoted. I don't think it's being promoted because it's got any scientific virtues. I think it's being promoted by people who've got a conservative, you know, moral religious perspective on life. I don't think it's a scientific issue. There are interesting scientific issues in this world, I mean, there surely are, about, let's say, the nature of human evolution or something like that, but this is not one of them.

Carlisle: Then against the backdrop, and you may dispute this, so let me give you all the room in the world—you'll take it anyway—so I don't have a problem there. But the way I want to frame this question is by the assumption that there is a backdrop of 2,500 years of design science that goes from Aristotle through Newton, and maybe up until the *Origin of Species*. So against this backdrop, what is it that necessitates that post-Darwinian science be nondesign science in nature?

Ruse: You have to be careful about the word *design*. I mean, I think Darwinian science takes very seriously the design metaphor, the notion that organisms are designed. But what Darwin does is show that you can explain it naturalistically, and so you don't have to invoke God as a hands-on worker.

Darwin himself certainly thought that God stood somewhere behind it all, and I take it that any Christian is going to believe that. I don't think any Christian is going to say that this no longer means that God is the designer. What it means is, we cannot prove the existence of God from design in the world. It's much the other way: that we look at the world through the filter of God and try to understand it.

If anybody looked at Auschwitz and said, "Ah-ha! This proves God," you'd have to be immoral. But it doesn't follow that a Christian cannot look at Auschwitz and say, "Let me try to understand and interpret this in a Christian sense." Are we going to talk in terms of kenosis, for instance, that God has voluntarily given up his powers at times like this? I take it that's what the Christian is going to do.

Carlisle: Okay, in the same vein, you've said that Intelligent Design is not science. Do you believe that Intelligent Design would have been perceived as nonscience by, say, Aristotle or Galileo or Newton?

Ruse: No, well, I don't know what Aristotle would have said. I mean, we're dealing with a different perspective. Probably Plato would have. I think that somebody like Paley would have said, you know, terrific, but he's living in the pre-Darwinian period.

Carlisle: So I guess the question is, what do you think happened by way of—or at the moment of—Darwinian theory?

Ruse: Darwin came along with a mechanism that could explain design, and that was the end of it. I mean that was why David Hume, for all that he roughed up the argument for design in dialogue, right at the end nevertheless says that it does seem that there's something there all along. He was right to say that. I agree with Dawkins that before Darwin it was not a wise position to be an atheist. Only after Darwin was it possible to be an intellectually fulfilled atheist. And I think that Dawkins was right. It doesn't say that you have to be an atheist, it means that it's possible to be one.

Carlisle: It seems to me that Dawkins essentially believes atheism to be a natural conclusion drawn from Darwinian theory.

Ruse: Of course I differ from Dawkins, but the point is, you see I think Dawkins was a sort of—well, let's just say it—village idiot in terms of what religion and atheism are.

If atheism is simply defined as saying, "I don't believe the Christian story so I am an atheist and I think everybody should be," then I think it's the only reasonable thing to be in the face of modern science. But I don't think it's quite that simple.

I think you can define Christianity, or certainly religious beliefs, in ways which state those sort of things, and you start to talk about God as transcendent, and you know, the holy other. As long as you think of God as being in some sense an old man with a beard and in a nightgown, then he's going to be vulnerable to science. But if you start to think of God as, in some sense, you know, the "Other," and that of which we cannot speak, and all those things that people like Barth and others have been saying, then I think you've got a, I mean, you may or may not believe in it, but I don't think it gives way to … to you know, quick refutations in quite the way that Dawkins argues.

I also beg to differ from Dawkins, Dennett, and others on the value of religion. I mean, they basically see religion as wholly evil, you know, what's that Weinberg statement: good people do good things and bad people do bad things, but it takes religion to get good people to do bad things. Well, yes, but I think there's been some people who have done some very good things because they're religious. What should I say, the Quakers and slavery, for instance, Elizabeth Fry [incidentally she is very famous in England] and her work in the prisons. I think these people did this, in many respects, because they were trying to serve their Lord.

Carlisle: Do you believe that science is science for all time, or that science evolves as a mode of knowledge?

Ruse: I don't think that science—I mean, I think it has done so in the sense of its not having been solidified in the last 300 or 400 years. But I don't see any movement in that in the last, say, 200 years or something like that. So if you're talking about science as a method, science as an enterprise, I don't see much change going on there.

Well, of course, you're going to get certain changes, such as whether or not it makes sense to talk in statistical terms, or things like that. But on the other hand, if you're talking about science, scientific theories, then basically of course, like anybody else I recognize that these are subject to change and evolution, although I'm not one who thinks that everything is up for grabs. It seems to me that the earth is not the center of the universe, and that organisms really did evolve.

I think there comes a point when certain discussions are finished.

Plate tectonics, the double helix, the DNA molecule is the carrier of information, and it's a double helix with four bases. It seems to me that those are never going to change. We may qualify them, but basically the fact that we've got 46 chromosomes, males and females have different sex chromosomes, you know, those it seems to me are established for all time. But on the other hand, things like Ed Wilson's theory of island biogeography, well maybe they'll never solve those.

Carlisle: Would you say then that as an epistemology, science is pretty much a sort of set approach to ….

Ruse: I just tried to tell you that, only I think it can be qualified. I mean it obviously needs to be qualified with the coming of things like quantum theory and statistics, and that sort of thing. But if you allow that kind of evolution, then I would say, yes, science is fixed once and for all. It is an enterprise that attempts to explain through unbroken law.

Carlisle: A friend and science writer has made the observation that neo-Darwinists seem to believe randomness applies to everyone but themselves. My question is, what do you make of the quandary that so-called products of randomness have given us?

Ruse: That seems to be one of those flip statements that journalists make. I'm not sure it's a particularly helpful statement. First of all, I'm not sure that as a neo-Darwinian, I'm that into randomness.

I'm certainly prepared to agree you can't tell when a particular mutation is going to occur rather than another, but I think on the other hand, you can quantify them. I don't think that they happen randomly in the sense that they're uncaused. I think you know what the causes are. I think that randomness is one of those relative concepts that you use in a certain context, as Darwin said, to point to ignorance.

If you go to an island off the coast of, let's say, South America, what do you expect to find? You expect to find South American-type forms. I don't think it's random. And if you don't, there's going to be a damn good reason, you know, currents or something like that. So I don't find these things particularly random at all. I don't think that Darwinism promotes randomness in that sort of sense. Certainly it may not promote progress, but things are going to go one way or the other depending on the circumstances. But I don't see that as random in any sort of cultural sort of tone.

Carlisle: But there does seem to be a problem—and I'm not sure that this is an argument at all that would be pursued by Intelligent Design people—when you have somebody like Dawkins talking about randomness in order to avoid some sort of purpose that might point to God.

Ruse: I think this is the thing: I wouldn't put one against the other. I think purpose or nonpurpose—of course if you don't think there's purpose you don't think it's necessarily random. I could think that the planets going around the sun, I don't think they've got a purpose, but I wouldn't say that their going around the sun is random. I would say that there's just no purpose.

I just expressed my own discomfort with terms like randomness. I don't think that you shouldn't use them, but I do think that you should recognize when they're being used and when to use them, and I think this is a point that Darwin himself makes.

Carlisle: There seems to be a desire for some kind of order insofar as that's what theories attempt to do. And yet, at the same time there does seem to be a desire for "randomness" to describe the way that nature operates.

Ruse: Again, I wouldn't use the word *random* there at all. I would say, of course, things are going to change depending on the way that, you know, external forces impinge upon them. And, of course, the external forces may vary disparately, but you know, I am uncomfortable using the word "random" at a time like that. Because it could just be that every year an island is going to be invaded, and I don't call that particularly random, although it may be that you can't tell which particular species are going to hit the island. Maybe they're random in that sense, but the response is not necessarily going to be random. I'm uncomfortable using the term random in that sort of bigger sense. I would certainly want to say I see evolution as being nonprogressive. I see evolution as not having some ultimate purpose or something like that, but I'm uncomfortable saying that evolution is random, because it seems to me that organisms, or humans, are anything but—what I mean by random is that randomness to me spells a mess, and humans are not random in that sense.

Randomness spells no way of figuring it out. It spells a random sequence of numbers or something like that. I see all of that, you know, that's how I would use the term *random*. I just don't think of humans as being particularly random. I think of them as being very sophisticated, if you like, biological machines which evolved in the face of certain needs, opportunities, and things of that nature. I don't see that as random at all. I don't see it as being necessarily progressive, because we might have gone other ways, and other ways might have been better: it's very expensive to maintain humans. We need a lot of protein, we need a lot of meat, so I mean, there are times when it's better to be not so expensive, so I don't see progress in that sense. I don't see purpose in that sense. I think it's perfectly open to you, as a Christian, to say I've made a Christian commitment and therefore I look at the world and I see God working his purpose out, and I see nothing inconsistent with somebody like you saying that. It's not a vision that I share, but it's not a vision I don't share because of how the world is. It's a vision I don't share because you're a person of faith and I am not.

Carlisle: So I guess insofar as that's the case, where do you see the fundamental sort of problem that makes you a person who is not of faith and me a person who is?

Ruse: I wouldn't use the word *problem* here, you see. I mean the point is, you, for whatever reason, maybe because of your background, or because of your psychological nature or maybe because God is speaking to you directly in some way, are a person of faith, and I am not.

It just doesn't mean anything to me. I also don't want to say things like, "Frankly I feel much more comfortable now that I'm not worried about the Presbyterian God that seemed to hover above me."

Like anybody who's been to boarding school, I think of God as being rather like my headmaster. My life is happier and more meaningful without religion hanging over me like a gray cloud.

Ultimately if I say, for whatever reason, you put it together one way and I put it together another way, and I mean, its like two different paradigms, we just don't share the same overlapping character. Of course, I would want to go on and draw certain theological consequences. I would be inclined to say, and therefore it's a bit mean of Paul to say, "You can't be saved unless you believe," but then if you were a Calvinist you'd probably come back and say, "Well, that's the way the cookie crumbles." You know, you're a goat and I'm a sheep, but you know, you can keep that up, but ultimately, as I say, "I put it down to faith."

Intelligent Design Timeline

13.7 billion years ago

The universe, as we understand it, is believed to have begun with the great cosmological explosion we've come to refer to as "the big bang."

Fifth century B.C.E.

The Biblical accounts of the world's beginnings are put into their final form. Commonly attributed to Moses, Genesis begins: "In the beginning God created the heavens and the earth. The earth was without form and void, and darkness was upon the face of the deep; and the Spirit of God was moving over the face of the waters."

Fourth century B.C.E.

Aristotle states that—though man may not actually see it—there appears to be purpose in nature. "It is absurd to suppose that purpose is not present because we do not observe the agent deliberating," he writes.

Aristotle develops a model for the geocentric (Earth-centered) universe, with the sun and planets rotating around the earth.

Third century C.E.

Plotinus, the father of neo-Platonism, teaches that there is a transcendent "One." The subsequent neo-Platonists held that all physical matter bore an animating life, or soul.

Thirteenth century

Thomas Aquinas, who believed that the existence of God could in fact be proved, completes his most famous work, *Summa Theologica*. In doing so, he incorporates Aristotle's "concept of purpose into Christian theology." Aquinas argues that nature cannot be fully explained without an "understanding of purpose."

Sixteenth century

Nicolas Copernicus suggests the sun is at the center of the universe.

Seventeenth century

Galileo defends heliocentrism. He has since become known as "the father of modern astronomy," "modern physics," and "modern science." As such, he figures into the modern debate over Intelligent Design.

1749–1844

Several pre-Darwin European naturalists—including Comte de Buffon, Erasmus Darwin, and Jean-Baptiste Lamarck—begin promoting the theory that "species are related by descent."

1759

Voltaire's *Candide* is published. This landmark Enlightenment work criticizes the idea that natural disasters are the work of divine intervention.

1802

William Paley publishes *Natural Theology*, in which he likens finding a watch in the woods to deducing design in the universe. To Paley, an intelligent designer of the infinitely complex universe was as believable as the designer of the detailed watch.

1847

Though considered to be essentially unrelated to the current Intelligent Design movement, the scientific term Intelligent Design is first published in an article in *Scientific American*.

1859

Charles Darwin's *Origin of Species* is published.

1859–1920s

Darwin's theory of evolution gains scientific credibility, and is subsequently applied as a social theory. Social Darwinists argue that the "fittest" in societies should be permitted to succeed, and the "unfit" allowed to decline. This thinking unfortunately spawns institutionalized concepts of a "superior race, and gives rise to eugenics."

1860

The Great Debate at Oxford University is held.

Bishop Samuel Wilberforce attacks Darwin's theory at a meeting of the British Association for the Advancement of Science. Scientists Thomas Huxley and Joseph Hooker defend Darwin's theory. Both parties claim victory.

1863

Thomas Huxley's *Evidence on Man's Place in Nature* is published.

1871

Darwin's *The Descent of Man* is published.

1917

Albert Einstein's general theory of relativity asserts that the universe is not a static space-time context, but is intrinsically dynamic.

1923

New Geology by George McCready Price is published. McCready, an early founder of the American creationist movement, says that the world's current geological features were a product of the biblical flood, which submerged all of life, save Noah, his family, and the animals on the ark. The presently visible sedimentary layers of the earth, he contends, were evidence of the flood, and that its fossil deposits reveal the organic life which existed at the time of the deluge.

1925

Lawyers William Jennings Bryan and Clarence Darrow go head to head in the now-famous Dayton, Tennessee, courtroom drama in which John Thomas Scopes (defended by Darrow) is tried for violating a state law banning the teaching of evolution. Scopes is found guilty and sentenced to a fine of $100.

1948

Religious teaching in public schools is banned by the United States Supreme Court, stating, "the First Amendment rests upon the premise that both religion and government can best work to achieve their lofty aims if each is left free from the other within its respective sphere."

1950

Astronomer Sir Fred Hoyle coins the term big bang, despite his own opposition to the theory.

1953

Primitive "laboratory atmosphere" is created by Stanley Miller and Harold C. Urey (Miller-Urey experiment), suggesting the possibility that organic life could come from inorganic matter.

The structure and significance of DNA, discovered in the nineteenth century, is first published by Francis Crick and James Watson.

1957

Geneticist J. B. S. Haldane's paper "The Cost of Natural Selection" questions the possibility that evolution could have evolved human beings from a common ancestor.

1961

The Genesis Flood by John C. Whitcomb Jr. and Henry M. Morris is published. The book gives birth to what today is known as "scientific creationism."

1968

The U.S. Supreme Court, stating the First Amendment must be protected and that the U.S. government must remain neutral on matters of religion, strikes down an anti-evolution law in Arkansas.

1973

Astrophysicist Brandon Carter introduces the "anthropic principle," which describes the apparently uncanny way in which the universe seems created for human existence.

1984

The Mystery of Life's Origins: Reassessing Current Theories, by Charles B. Thaxton, considered in some circles to be the first major work in the modern Intelligent Design movement, is published.

1986

Richard Dawkins's *The Blind Watchmaker*, which directly criticizes Paley's watchmaker theory, is published.

1987

The U.S. Supreme Court says that any law requiring the teaching of creationism in schools is in violation of the First Amendment.

1989

Of Pandas and People: The Central Question of Biological Origins, by Percival Davis and Dean Kenyon, is published.

1990

The Discovery Institute, the Intelligent Design movement's think tank, is founded in Seattle, Washington.

1991

Darwin on Trial, by Phillip E. Johnson, is published.

The book, which coins the term intelligent design, becomes an essential handbook for the Intelligent Design movement.

Gallup poll shows 90 percent of Americans polled believe God created the world.

1992

The Wedge Strategy—outlining the various goals and objectives of the Intelligent Design movement—is first published by the Discovery Institute's Center for the Renewal of Science & Culture.

1996

Pope John Paul II proclaims evolution to be "more than a hypothesis" and says that there is no "essential contradiction between evolutionary science and Catholicism."

Phillip E. Johnson becomes a founding adviser for the establishment of the Discovery Institute's Center for Science and Culture (formerly known as the Center for the Renewal of Science and Culture).

Darwin's Black Box, by Michael Behe, is published.

1998

The Design Inference: Eliminating Chance Through Small Probabilities, by William Dembski, is published.

1999

Dembski becomes head of Baylor University's Center for the Study of Intelligent Design.

2001

Phillip E. Johnson assists in the drafting of the Santorum Amendment—proposed by U.S. Senator Rick Santorum (R-PA) to what became the "No Child Left Behind Act." The amendment promotes the teaching of Intelligent Design. Though ultimately removed from the bill, proponents of ID view the effort as a victory.

A Zogby International poll finds that most American adults favor teaching arguments against evolution.

2002

Hearings are held in Ohio on what should be taught regarding evolution. ID is ultimately rejected.

2004

Atheist Antony Flew sends a letter to the editor of *Philosophy Now* suggesting that his atheistic views might be changing. According to Flew, "[Darwin probably] believed that life was miraculously breathed into that primordial form of not always consistently reproducing life by God, though not the revealed God of then contemporary Christianity, who had predestined so many of Darwin's friends and family to an eternity of extreme torture."

Baylor University's Intelligent Design center is disbanded.

An article by Stephen Meyer, director of the Discovery Institute's Center for Science and Culture, is published in *Proceedings of the Biological Society* of Washington. Meyer's piece is the first paper on Intelligent Design to appear in a "peer-reviewed journal." The magazine's editor is criticized by the scientific community.

In November, a CBS News poll reveals that more than half of Americans surveyed believe God created human beings in their present form.

That same month, a Gallup poll reveals that only 13 percent of the poll participants believe that God had no part in the evolution or creation of human life, whereas 38 percent say they believe humans evolved from less-advanced forms, with God guiding the process.

2005

Antony Flew authors a new introduction to *God & Philosophy*, stating that he will "follow the argument wherever it leads."

November 2005, the State Board of Education in Kansas passes a measure similar to that of Ohio in 2002.

Sources: This Intelligent Design Timeline is drawn from a variety of sources, including this book's chapters as well as timelines and articles published by *Science & Technology News*, *USA Today*, NASA, *Indianapolis Star*, and the American Museum of Natural History.

The Wedge Project

The Wedge Project (also known as the Wedge Strategy, the Wedge Document, or The Wedge) is a paper first published in 1992 by The Discovery Institute's Center for the Renewal of Science & Culture.

The Wedge Document outlines various goals and objectives of the Intelligent Design movement, and it has been widely circulated and re-published on various Internet sites since it was first introduced to the general public. The Document has been praised by proponents of Intelligent Design and roundly criticized by ID's opponents.

THE WEDGE STRATEGY
BY THE
CENTER FOR THE RENEWAL OF SCIENCE & CULTURE

INTRODUCTION

The proposition that human beings are created in the image of God is one of the bedrock principles on which Western civilization was built. Its influence can be detected in most, if not all, of the West's greatest achievements, including representative democracy, human rights, free enterprise, and progress in the arts and sciences.

Yet a little over a century ago, this cardinal idea came under wholesale attack by intellectuals drawing on the discoveries of modern science. Debunking the traditional conceptions of both God and man, thinkers such as Charles Darwin, Karl Marx, and Sigmund Freud portrayed humans not as moral and spiritual beings, but as animals or machines who inhabited a universe ruled by purely impersonal forces and whose behavior and very thoughts were dictated by the unbending forces of biology, chemistry, and environment.

This materialistic conception of reality eventually infected virtually every area of our culture, from politics and economics to literature and art.

The cultural consequences of this triumph of materialism were devastating. Materialists denied the existence of objective moral standards, claiming that environment dictates our behavior and beliefs. Such moral relativism was uncritically adopted by much of the social sciences, and it still under-girds much of modern economics, political science, psychology and sociology.

Materialists also undermined personal responsibility by asserting that human thoughts and behaviors are dictated by our biology and environment. The results can be seen in modern approaches to criminal justice, product liability, and welfare. In the materialist scheme of things, everyone is a victim and no one can be held accountable for his or her actions.

Finally, materialism spawned a virulent strain of utopianism. Thinking they could engineer the perfect society through the application of scientific knowledge, materialist reformers advocated coercive government programs that falsely promised to create heaven on earth.

Discovery Institute's Center for the Renewal of Science and Culture seeks nothing less than the overthrow of materialism and its cultural legacies. Bringing together leading scholars from the natural sciences and those from the humanities and social sciences, the Center explores how new developments in biology, physics and cognitive science raise serious doubts about scientific materialism and have re-opened the case for a broadly theistic understanding of nature. The Center awards fellowships for original research, holds conferences, and briefs policymakers about the opportunities for life after materialism.

The Center is directed by Discovery Senior Fellow Dr. Stephen Meyer. An Associate Professor of Philosophy at Whitworth College, Dr. Meyer holds a Ph.D. in the History and Philosophy of Science from Cambridge University. He formerly worked as a geophysicist for the Atlantic Richfield Company.

THE WEDGE STRATEGY

Phase I
Scientific Research, Writing & Publicity

Phase II
Publicity & Opinion-making

Phase III
Cultural Confrontation & Renewal

THE WEDGE PROJECTS

Phase I
Scientific Research, Writing & Publication

- Individual Research Fellowship Program
- Paleontology Research program (Dr. Paul Chien et al.)
- Molecular Biology Research Program (Dr. Douglas Axe et al.)

Phase II

- Publicity & Opinion-making
- Book Publicity
- Opinion-Maker Conferences
- Apologetics Seminars
- Teacher Training Program
- Op-ed Fellow
- PBS (or other TV) Co-production
- Publicity Materials / Publications

Phase III
Cultural Confrontation & Renewal

- Academic and Scientific Challenge Conferences
- Potential Legal Action for Teacher Training
- Research Fellowship Program: shift to social sciences and humanities

FIVE YEAR STRATEGIC PLAN SUMMARY

The social consequences of materialism have been devastating. As symptoms, those consequences are certainly worth treating. However, we are convinced that in order to defeat materialism, we must cut it off at its source. That source is scientific material-ism. This is precisely our strategy. If we view the predominant materialistic science as a giant tree, our strategy is intended to function as a "wedge" that, while rela-tively small, can split the trunk when applied at its weakest points. The very begin-ning of this strategy, the "thin edge of the wedge," was Phillip Johnson's critique of Darwinism begun in 1991 in Darwinism on Trial, and continued in Reason in the Balance and Defeating Darwinism by Opening Minds. Michael Behe's highly success-ful Darwin's Black Box followed Johnson's work. We are building on this momentum, broadening the wedge with a positive scientific alternative to materialistic scientific theories, which has come to be called the theory of intelligent design (ID). Design theory promises to reverse the stifling dominance of the materialist worldview, and to replace it with a science consonant with Christian and theistic convictions.

The Wedge strategy can be divided into three distinct but interdependent phases, which are roughly but not strictly chronological. We believe that, with adequate sup-port, we can accomplish many of the objectives of Phases I and II in the next five years (1999-2003), and begin Phase III (See "Goals/Five Year Objectives/Activities").

Phase I: Research, Writing and Publication
Phase II: Publicity and Opinion-making
Phase III: Cultural Confrontation and Renewal

Phase I is the essential component of everything that comes afterward. Without solid scholarship, research and argument, the project would be just another attempt to indoctrinate instead of persuade. A lesson we have learned from the history of science is that it is unnecessary to outnumber the opposing establishment. Scientific revolu-tions are usually staged by an initially small and relatively young group of scientists who are not blinded by the prevailing prejudices and who are able to do creative work at the pressure points, that is, on those critical issues upon which whole systems of thought hinge. So, in Phase I we are supporting vital witing and research at the sites most likely to crack the materialist edifice.

Phase II. The primary purpose of Phase II is to prepare the popular reception of our ideas. The best and truest research can languish unread and unused unless it is prop-erly publicized. For this reason we seek to cultivate and convince influential individu-als in print and broadcast media, as well as think tank leaders, scientists and academics, congressional staff, talk show hosts, college and seminary presidents and faculty, future talent and potential academic allies. Because of his long tenure in politics, journalism and public policy, Discovery President Bruce Chapman brings to the project rare knowledge and acquaintance of key op-ed writers, journalists, and political leaders. This combination of scientific and scholarly expertise and media and political connections makes the Wedge unique, and also prevents it from being "merely academic." Other activities include production of a PBS documentary on intelligent design and its implications, and popular op-ed publishing. Alongside a focus on influential opinion-makers, we also seek to build up a popular base of support among our natural constituency, namely, Christians. We will do this primarily through apologetics seminars. We intend these to encourage and equip believers with new scientific evidence's that support the faith, as well as to "popularize" our ideas in the broader culture.

Phase III. Once our research and writing have had time to mature, and the public prepared for the reception of design theory, we will move toward direct confrontation with the advocates of materialist science through challenge conferences in significant academic settings. We will also pursue possible legal assistance in response to resistance to the integration of design theory into public school science curricula. The attention, publicity, and influence of design theory should draw scientific materialists into open debate with design theorists, and we will be ready. With an added emphasis to the social sciences and humanities, we will begin to address the specific social consequences of materialism and the Darwinist theory that supports it in the sciences.

GOALS

Governing Goals

- ◆ To defeat scientific materialism and its destructive moral, cultural and political legacies.

- ◆ To replace materialistic explanations with the theistic understanding that nature and human beings are created by God.

Five Year Goals

- ◆ To see intelligent design theory as an accepted alternative in the sciences and scientific research being done from the perspective of design theory.

- ◆ To see the beginning of the influence of design theory in spheres other than natural science.

- ◆ To see major new debates in education, life issues, legal and personal responsibility pushed to the front of the national agenda.

Twenty Year Goals

- To see intelligent design theory as the dominant perspective in science.

- To see design theory application in specific fields, including molecular biology, biochemistry, paleontology, physics and cosmology in the natural sciences, psychology, ethics, politics, theology and philosophy in the humanities; to see its influence in the fine arts.

- To see design theory permeate our religious, cultural, moral and political life.

FIVE YEAR OBJECTIVES

1. A major public debate between design theorists and Darwinists (by 2003)

2. Thirty published books on design and its cultural implications (sex, gender issues, medicine, law, and religion)

3. One hundred scientific, academic and technical articles by our fellows

4. Significant coverage in national media:

   - Cover story on major news magazine such as Time or Newsweek.

   - PBS shows such as Nova treating design theory fairly.

   - Regular press coverage on developments in design theory.

   - Favorable op-ed pieces and columns on the design movement by 3rd party media.

5. Spiritual & cultural renewal:

   - Mainline renewal movements begin to appropriate insights from design theory, and to repudiate theologies influenced by materialism.

   - Major Christian denomination(s) defend(s) traditional doctrine of creation & repudiate(s).

   - Darwinism Seminaries increasingly recognize & repudiate naturalistic presuppositions.

   - Positive uptake in public opinion polls on issues such as sexuality, abortion and belief in God.

6. Ten states begin to rectify ideological imbalance in their science curricula & include design theory

7. Scientific achievements:

◆ An active design movement in Israel, the UK and other influential countries outside the US

◆ Ten CRSC Fellows teaching at major universities

◆ Two universities where design theory has become the dominant view

◆ Design becomes a key concept in the social sciences Legal reform movements base legislative proposals on design theory

ACTIVITIES

(1) Research Fellowship Program (for writing and publishing)

(2) Front line research funding at the "pressure points" (e.g., Paul Chien's Chengjiang Cambrian Fossil Find in paleontology, and Doug Axe's research laboratory in molecular biology)

(3) Teacher training

(4) Academic Conferences

(5) Opinion-maker Events & Conferences

(6) Alliance-building, recruitment of future scientists and leaders, and strategic partnerships with think tanks, social advocacy groups, educational organizations and institutions, churches, religious groups, foundations and media outlets

(7) Apologetics seminars and public speaking

(8) Op-ed and popular writing

(9) Documentaries and other media productions

(10) Academic debates

(11) Fund Raising and Development

(12) General Administrative support

THE WEDGE STRATEGY PROGRESS SUMMARY

Books

William Dembski and Paul Nelson, two CRSC Fellows, will very soon have books published by major secular university publishers, Cambridge University Press and The University of Chicago Press, respectively. (One critiques Darwinian materialism; the other offers a powerful alternative.)

Nelson's book, On Common Descent, is the seventeenth book in the prestigious University of Chicago "Evolutionary Monographs" series and the first to critique neo-Darwinism. Dembski's book, The Design Inference, was back-ordered in June, two months prior to its release date.

These books follow hard on the heals of Michael Behe's Darwin's Black Box (The Free Press) which is now in paperback after nine print runs in hard cover. So far it has been translated into six foreign languages. The success of his book has led to other secular publishers such as McGraw Hill requesting future titles from us. This is a break-through.

InterVarsity will publish our large anthology, Mere Creation (based upon the Mere Creation conference) this fall, and Zondervan is publishing Maker of Heaven and Earth: Three Views of the Creation-Evolution Controversy, edited by fellows John Mark Reynolds and J.P. Moreland.

McGraw Hill solicited an expedited proposal from Meyer, Dembski and Nelson on their book Uncommmon Descent. Finally, Discovery Fellow Ed Larson has won the Pulitzer Prize for Summer for the Gods, his retelling of the Scopes Trial, and InterVarsity has just published his co-authored attack on assisted suicide, A Different Death.

Academic Articles

Our fellows recently have been featured or published articles in major scientific and academic journals in The Proceedings to the National Academy of Sciences, Nature, The Scientist, The American Biology Teacher, Biochemical and Biophysical Research Communications, Biochemistry, Philosophy and Biology, Faith & Philosophy, American Philosophical Quarterly, Rhetoric & Public Affairs, Analysis, Book & Culture, Ethics & Medicine, Zygon, Perspectives on Science and the Christian Faith, Religious Studies, Christian Scholars' Review, The Southern Journal of Philosophy, and the Journal of Psychology and Theology. Many more such articles are now in press or awaiting review at major secular journals as a result of our first round of research fellowships. Our own journal, Origins & Design, continues to feature schol-arly contributions from CRSC Fellows and other scientists.

Television and Radio Appearances

During 1997 our fellows appeared on numerous radio programs (both Christian and secular) and five nationally televised programs, TechnoPolitics, Hardball with Chris Matthews, Inside the Law, Freedom Speaks, and Firing Line. The special edition of

TechnoPolitics that we produced with PBS in November elicited such an unprecedented audience response that the producer Neil Freeman decided to air a second episode from the "out takes." His enthusiasm for our intellectual agenda helped stimulate a special edition of William F. Buckley's Firing Line, featuring Phillip Johnson and two of our fellows, Michael Behe and David Berlinski. At Ed Atsinger's invitation, Phil Johnson and Steve Meyer addressed Salem Communications' Talk Show Host conference in Dallas last November. As a result, Phil and Steve have been interviewed several times on Salem talk shows across the country. For example, in July Steve Meyer and Mike Behe were interviewed for two hours on the nationally broadcast radio show Janet Parshall's America. Canadian Public Radio (CBC) recently featured Steve Meyer on their Tapestry program. The episode, "God & the Scientists," has aired all across Canada. And in April, William Craig debated Oxford atheist Peter Atkins in Atlanta before a large audience (moderated by William F. Buckley), which was broadcast live via satellite link, local radio, and Internet "webcast."

Newspaper and Magazine Articles

The Firing Line debate generated positive press coverage for our movement in, of all places, The New York Times, as well as a column by Bill Buckley. In addition, our fellows have published recent articles & op-eds in both the secular and Christian press, including, for example, The Wall Street Journal, The New York Times, The Washington Times, National Review, Commentary, Touchstone, The Detroit News, The Boston Review, The Seattle Post-Intelligencer, Christianity Today, Cosmic Pursuits, and World. An op-ed piece by Jonathan Wells and Steve Meyer is awaiting publication in the Washington Post. Their article criticizes the National Academy of Science book Teaching about Evolution for its selective and ideological presentation of scientific evidence. Similar articles are in the works.

Further Reading

Behe, Michael J. *Darwin's Black Box*. New York: The Free Press, 1996.

Darwin, Charles. *Origin of Species*. New York: New York University Press, 1988.

Dawkins, Richard. *The Blind Watchmaker*. New York: W.W. Norton, 1996.

Dembski, William. *Mere Creation*. Downers Grove, IL: InterVarsity Press, 1998.

Dennett, Daniel. *Darwin's Dangerous Idea*. New York: Simon and Schuster, 1995.

Denton, Michael J. *Nature's Destiny*. New York: The Free Press, 1998.

Einstein, Albert. *Relativity*. New York: Crown Publishers, 1961.

Gould, Stephen Jay. *Rocks of Ages*. New York: Ballantine Publishing, 1999.

Hawking, Stephen. *A Brief History of Time*. New York: Bantam Books, 1988.

Hooper, Judith. *Of Moths and Men*. New York: W.W. Norton, 2002.

Johnson, Phillip. *Darwin on Trial*. Washington, DC: Regnery Gateway, 1991.

Miller, Kenneth. *Finding Darwin's God*. New York: HarperCollins, 1999.

Pearcey, Nancy R., and Charles B. Thaxton. *The Soul of Science: Christian Faith and Natural Philosophy*. Wheaton, IL: Crossway Books, 1994.

Paley, William. *Natural Theology*. New York: Oxford University Press, 1828.

Polkinghorne, John. *Belief in God in an Age of Science*. New Haven, CT: Yale University Press, 1998.

Ruse, Michael. *The Evolution-Creation Struggle*. Cambridge, MA: Harvard University Press, 2005.

Schroeder, Gerald L. *The Science of God*. New York: The Free Press, 1997.

Sobel, Dava. *Galileo's Daughter*. New York: Walker Publishing, 1999.

Teresi, Dick. *Lost Discoveries*. New York: Simon and Schuster, 2002.

Wilson, E. O. *On Human Nature*. Cambridge, MA: Harvard University Press, 1998.

Recommended Websites

Mainstream Science

American Institute of Biological Sciences
www.actionbioscience.org

The Talk Origins Archive
www.talkorigins.org

Kenneth Miller's mainstream biology website
www.millerandlevine.com

Stephen Jay Gould's evolution website
www.stephenjaygould.org

Evolution website for all ages
www.evolution.berkeley.edu

Website supporting secularism and naturalism
www.infidels.org

Intelligent Design

The Discovery Institute
www.discovery.org

Intelligent Design and Evolution Awareness Center
www.ideacenter.org.

William Dembski's website
www.designinference.com

Access Research Network
www.arn.org

"ID Net" website
www.intelligentdesignnetwork.org

Index

D

E

F

social science, 67-68
Social Darwinism, 69
survival of the fittest, 68-69
sociobiology
The New Synthesis, 181
Wilson, Edward Osborne, 181-182
soul, life's origins, 98
specified complexity, Intelligent Design biology, 145
everyday use, 146
nothing is random, 145-146
objectivity, 147
Spencer, Herbert, 69
spontaneous life
mass of mud, 99
Miller-Urey experiment, 100
progress at molecular level, 100-101
Stear, John, *A Critical Look at Creationist Paleontology*, 114
Stenger, Victor J., 94
sticky white blood cells, blood clotting system, 109
stratosphere, 97
supernatural, theology study, 17-18
survival fittest, 68-69

T

teleology, 49
Teresi, Dick, *Lost Discoveries*, 14, 37
testing, hypotheses, 61
Thaxton, Charles, 10, 78
abductive inference, 122-123
The Soul of Science, 40
theology, 17
failure to meet scientific criteria, 18
miracles, 18-19
supernatural, 17-18
thermodynamic law, Intelligent Design physics
entropy, 124
movement towards disorder, 125-126
organization of chaos, 125

thermosphere, 97
think tanks, Discovery Institute, 10
Tillich, Paul, 76
transcendents, 62
troposphere, 97
truth, nature of understanding, 218-219
TTSS (type three secretory system), 107

U

unity of knowledge, 248
movement towards, 250
mutual poverty, 249-250
years of interrelation, 249
universal acid, 235
universal common descent, biology, 112-113
universe
big bang theory, 84-85
fine-tuned, 90, 126
firing squad analogy, 126-127
humans in center, 127
millions of possibilities, 90-91
multiverse, 91, 127-128
Urey, Harold C., 100

V-W

violin model, 6

Wallace, Alfred Russel, 168
water
anthropic principle, 89
chemistry, 96
Intelligent Design chemistry, 132
Watson, James, 69, 101
Wedge Document, 209
Wedge Project, 211
Wells, Dr. Jonathan, *Icons of Evolution*, 31
Whitcomb, John, *The Genesis Flood*, 24
White, Ellen G., 9
Wilkins, John, "Macroevolution," 112

X-Y-Z